INTEGRATION OF SCIENCE AND THE SYSTEMS APPROACH

CZECHOSLOVAK ACADEMY OF SCIENCES

Scientific Editor: Doc. dr. Karel Berka, DrSc.
Reviewer: Prof. dr. František Čížek, DrSc.

Integration of Science and the Systems Approach

edited by
ARKADIY DMITRIYEVICH URSUL
Department for Philosophy and Law
of the Academy of Sciences of the MSSR,
Kishinev
ZDENĚK JAVŮREK, JIŘÍ ZEMAN
Institute for Philosophy and Sociology
of the Czechoslovak Academy of Sciences,
Prague

ELSEVIER
Amsterdam—Oxford—New York—Tokyo 1984

Scientific Editor: Doc. Dr. Karel Berka, DrSc.
Reviewer: Doc. Dr. František Čížek, DrSc.

Published in co-edition with
ACADEMIA, Publishing House of the Czechoslovak Academy of Sciences, Prague

Distribution of this book is being handled by the following publishers

for the U. S. A. and Canada
Elsevier Science Publishing Company, Inc.
52 Vanderbilt Avenue
New York, N. Y. 10017 U. S. A.

for the East European Countries, China, North Korea, Cuba, Vietnam and Mongolia
Academia, Publishing House of the Czechoslovak Academy of Sciences, Prague

for all remaining areas
Elsevier Science Publishers,
Molenwerf 1
P. O. Box 211, 1000 AE Amsterdam, The Netherlands

Library of Congress Cataloging in Publication Data
Main entry under title:

Integration of science and the systems approach.

„Published in co-edition with Academia, Publishing House of the Czechoslovak Academy of Sciences, Prague" — T. p. verso.
Bibliography: p.
1. Science-Philosophy. 2. System theory. 3. Cognition. 4. Dialectical materialism. I. Javůrek, Zdeněk. II. Ursul, Arkadii Dmitrievich. III Zeman, Jiří.

Q175.15135 1984 501 82-20657
ISBN 0-444-99633-8

Printed in Czechoslovakia

Contents

LIST OF AUTHORS

PETR NIKOLAYEVICH FEDOSEYEV, Academician, Vice-President of the Academy of Sciences of the USSR, Moscow

HERBERT HÖRZ, Academician, Vice-Director of the Central Institute of Philosophy, Academy of Sciences of the German Democratic Republic, Berlin

ZDENĚK JAVŮREK, PhDr., CSc., scientific collaborator of the Institute for Philosophy and Sociology of the Czechoslovak Academy of Sciences, Prague

JANA PURŠOVÁ, PhDr., CSc., scientific collaborator of the Institute for Philosophy and Sociology of the Czechoslovak Academy of Sciences, Prague

GEORGIY IVANOVICH RUZAVIN, PhDr., DSc., scientific collaborator of the Philosophical Institute of the Academy of Sciences of the USSR, Moscow

VADIM NIKOLAYEVICH SADOVSKY, PhDr., DSc., Chief of the Laboratory VNIISI KGNT, Academy of Sciences of the USSR, Moscow

STANISLAV NIKOLAYEVICH SMIRNOV, PhDr., DSc., Chief of the Sector INION, Academy of Sciences of the USSR; Scientific Secretary of the Council on International Collaboration of the Academy of Sciences of the USSR, Moscow

JIŘINA STACHOVÁ, PhDr., CSc., scientific collaborator of the Institute for Philosophy and Sociology of the Czechoslovak Academy of Sciences, Prague

VIKTOR STEPANOVICH TYUKHTIN, PhDr., DSc., scientific collaborator of the Philosophical Institute of the Academy of Sciences of the USSR, Moscow

BORIS SERGEYEVICH UKRAINTSEV, PhDr., DSc., scientific collaborator of the Philosophical Institute of the Academy of Sciences of the USSR, Moscow

ARKADIY DMITRIYEVICH URSUL, PhDr., DSc., Chief of the Sector of the Department for Philosophy and Law of the Academy of Sciences of the Moldavian SSR, Kishinev

IVO ZAPLETAL, PhDr., CSc., scientific collaborator of the Institute for Philosophy and Sociology of the Czechoslovak Academy of Sciences, Prague

JINDŘICH ZELENÝ, Corresponding Member of the Czechoslovak Academy of Sciences, Vice-Director of the Institute for Philosophy and Sociology of the Czechoslovak Academy of Sciences, Prague

JIŘÍ ZEMAN, PhDr., DrSc., scientific collaborator of the Institute for Philosophy and Sociology of the Czechoslovak Academy of Sciences, Prague

Preface

The international collection of papers on "Integration of Science and the Systems Approach" is a publication which summarizes the research into topical problems concerning integration processes in scientific cognition in relation to systems approach; it includes contributions by outstanding philosophers and scientists from the USSR, the ČSSR and the GDR. As for its contents, it is linked to several preceding publications of international and interdisciplinary character, i.e. the collections of papers on "Time in Science and Philosophy", Elsevier—Academia, Amsterdam—Prague 1971 (ed. J. Zeman), "Entropy and Information in Science and Philosophy", Elsevier—Academia, Amsterdam—Prague 1975 (eds. L. Kubát—J. Zeman) and "Dialectics and Systems Approach", Svoboda, Prague 1979 (eds. Z. Javůrek—J. Zeman, in Czech).

It was the central idea of the collection of papers on time to trace out a certain field of problems in which interests of different disciplines such as physics, astronomy, geology, biology, psychology and philosophy meet and in which it is therefore possible to look for a certain unifying moment of these disciplines from the thematic and topical as well as formal and linguistic viewpoints. This idea was re-assumed by the following collection of papers on entropy and information which took up the endeavours to find common problems, concepts and approaches and was primarily based on the initial themes of cybernetics, above all the concept of entropy and information. The publication "Dialectics and Systems Approach" aimed at pointing out the connection between materialistic dialectics and systems thinking which represents, in a certain sense, a more general approach than is represented by the mere approach of cybernetics, while finding in the concept of system one of the central concepts penetrating all scientific disciplines and simultaneously subordinate to the conception of dialectical-materialistic monism. The present publication develops further the idea of systems approach, i.e. in a more general connection with integration trends in scientific cognition, with integration and unification functions and processes in science in general as well as in individual scientific disciplines. The problem of the world's material unity and the unity of scientific cognition is of a considerable theoretical and practical-methodological importance. Apart from the increasing differentiation of scientific cognition, the desirable process of their integration in the overall world outlook and practical attitudes is under way. New forms of interaction between the natural, social and technical sciences emerge and develop. Integration processes are also connected with the creation of new functions and a new structure of general scientific cognition. This complex and global problem has its important theoretical, methodological and philosophical aspects. This integration of science is closely connected with the development of systems thinking which aims at a general and concrete definition of the concept of system in scientific cognition, the definition of other related concepts and the elaboration of systems methods in science.

The book presented is divided into two parts — one being more general, the other more special. Part I, entitled "Integrative Functions in Science", includes the contributions by Soviet scientists P. N. Fedoseyev, A. D. Ursul, S. N. Smirnov, the German scientist H. Hörz and Czech scientists J. Zelený, J. Zeman and I. Zapletal and considers the general character of integration of scientific cognition in connection with materialistic dialectics and materialistic monism. Part II, entitled "Integration Processes in Science and Systems Approach", encompasses the contributions by Soviet scientists V. N. Sadovsky, G. I. Ruzavin, V. S. Tyukhtin, B. S. Ukrayintsev and Czech scientists Z. Javůrek, J. Puršová and J. Stachová; it follows the problems of integration processes in individual sciences and their development in connection with the development of systems approach. The editors of this collection of papers were led by the central idea of presenting an as far as possible complex outline of the given range of problems which would contribute to bring about desirable and necessary unifying trends in scientific cognition.

Z. Javůrek
A. D. Ursul
J. Zeman

I.
INTEGRATIVE FUNCTIONS OF SCIENCE

1. Philosophy and the Integration of Knowledge

P. N. FEDOSEYEV

The problem of the role and place held by philosophy in the system of the scientific cognition of objective reality is now becoming the subject of ever more heated discussions and controversies. One of the contributing factors in the mounting battle of ideas around the interconnection between philosophy and specific sciences is the development of integration processes in modern science, in particular, at the junctions of the social, natural, and technical branches of knowledge.

It is apparent that progress in science is being accompanied by the increasing ramification and the simultaneous mutual penetration of various branches of scientific knowledge, their synthesis and integration. In other words, the inner development of science which poses ever broader tasks in scientific quest, stimulates a desire to go beyond the boundaries of each branch of knowledge, which became isolated earlier, and leads to a growing interconnection between the natural, social and technical sciences. However, there is more to it than that. At present, the growing unity between the three main branches of scientific knowledge stems from the comprehensive nature of the problems whose solution is needed by social development. It is fair to say that it is impossible to solve the present-day genuinely tremendous social and economic problems facing mankind without the increasing interconnection and harmonious development of the natural, social and technical sciences.

Naturally, all this makes it even more urgent to elaborate the nature, character, forms, and boundaries of the current integration processes in a philosophical manner. The problem of the self-determination of philosophy is becoming particularly acute both from the viewpoint of the realization of its organic involvement in this law-governed movement of the scientific knowledge of the world and from the standpoint of its specific functions contributing to the materialization of the integrative tendencies in modern science.

It is Marxist-Leninist philosophy that offers a profound theoretical substantion for the law-governed patterns in the integration of scientific knowledge. Marx and Engels, proceeding from the principles of materialistic dialectics, foresaw the inevitable increase in the integration processes even at a time when the tendency for the growing differentiation of sciences still prevailed in the development of human cognition. Marx used to point to the historically conditioned and transient division of the natural and social sciences. He once observed that, figuratively speaking, with time "natural science will come to include the science of man to the same extent to which the science of man will include natural science: it will be *one* science".[1]

Engels, assessing the state of science in his day and, in particular, that of classical natural science, remarked that its successes and achievements were largely determined by the wide use of the analytical methods and procedures of research, which make it possible to single out each object and phenomenon of the surrounding reality and to study them separately. At the same time, Engels emphasized that the unilaterally absolutized analytical methods were increasingly showing their limitations. Metaphysical thinking had put up barriers between the individual forms of the motion of matter and the scientific disciplines engaged in their study which classical natural science was unable to overcome.

In the meantime the development of science made it possible to realize the relativity of these barriers more and more. A philosophical analysis of such fundamental achievements of nineteenth century natural science as the law of the conservation and the conversion of energy, cellular physiology, and the theory of biological evolution enabled Engels to formulate and thoroughly substantiate the principle of the unity of the world and the unity of human knowledge.

In 1878 Engels presented consistent and authentic evidence showing the material foundation of the unity of the world and its cognition in his famous work Anti-Dühring, the centenary of which was widely marked not so long ago. "The real unity of the world consists in its material nature," Engels observed. This is proved by the "long and difficult progress of philosophy and natural science".[2] The subsequent development of science not only brought fresh proof of the material unity of the world, but also came to rely on this principle as its world-viewing and methodological prerequisite. It became vital to adopt a dialectical method of thinking oriented to the universal connection between the phenomena and processes of objective reality, obtaining synthetic knowledge of objects, and broad theoretical generalizations overstepping the established boundaries between the disciplines. This method was needed if the barriers between the individual branches of science were to be overcome successfully, an all-round, integral view of the objects studied worked out. In this connection Engels underlined the particularly important and promising nature of the research conducted at the junctions of sciences.

"The whole of nature that is accessible to us," Engels observed, "forms a system, an aggregate connection of bodies" which influence one another, and this influence is nothing else but motion. Man cognized the universe as a mutual connection between bodies. Hence the logical conclusion that the integration of human knowledge was inevitable. The growing interconnection of the sciences reflected the real interconnection of phenomena in reality itself. The increasing integration of knowledge indicated that science was penetrating deeper and deeper into the dialectics of objective reality, the dialectics of nature and social development.

The subsequent development of science fully confirmed Engels' ideas. A synthesis of the notions of electricity and magnetism, the corpuscular properties of light, the development of physical chemistry and chemical physics, a synthesis of genetics and Darwin's theory of evolution, the study of the molecular and submolecular mechanisms of heredity, the drawing together of micro- and macrophysics in the present-day cosmological and astrophysical studies, the study of man at the junction of the social and biological sciences — these are just a few of the landmarks in the development of scientific cognition. They mean that it has attained a new level and is penetrating the objective dialectics of the development of reality more and more deeply.

The further development of a dialectical materialistic approach towards the interconnection of the sciences is contained in Lenin's works. Even in his early works aimed at the subjective sociology of populism, Lenin emphasized Marx's tenet of the natural historical necessity manifesting itself in the development of society. This tenet constitutes the basis of the dialectical materialistic methodology of cognizing social phenomena; its conscious and consistent application enabled Marx to discover objective law-governed patterns in the development of society and thus to place sociology on a scientific foundation for the first time. Incidentally, Lenin pointed to the closeness of Marx's dialectical-materialistic methodology based on social science material and the methodological premises contained in Darwin's theory of evolution studying the law-governed patterns of the development of organic nature.

The methodology of classical science developed under the strong influence of the philosophical concepts absolutizing the opposition between nature and spirit, matter and consciousness. Naturally enough, these concepts hampered the interaction of humanitarian and natural scientific cognition in studying many fundamental problems. In his work "Materialism and Empiriocriticism" Lenin proved that this opposition had an absolute value only as part of the principal epistemological question beyond which the relativity of this contraposition was undoubted. This premise cleared the way for materialistic and strictly scientific research within the concrete sciences into the problems arising at the junction of natural scientific and humanitarian cognition. This conclusion as well as other fundamental principles of Lenin's Materialism and Empiriocriticism related to the philosophical, dialectical-materialistic substantiation and analysis of the most important discoveries in physics at the turn of the century, were further developed in the idea of an alliance between philosophy and natural science.

Noting the fruitful influence of the sciences of nature on the development of the social sciences, Lenin observed: "It is common knowledge that a powerful current flowed from natural to social science not only in Petty's time, but in Marx's time as well. And this current remains just as powerful, if not more so, in the twentieth century, too."[3] At the same time, Lenin emphasized the methodological importance of philosophy and its significance from the point of view of Weltanschauung and the social sciences for the development of natural science. He called attention to the fact that it was as a result of the upheaval taking place in contemporary natural science that the main reactionary philosophical schools and the minor ones, the major and minor trends, were born as often as not. "It must be realized," Lenin pointed out, "that no natural science and no materialism can hold its own in the struggle against the onslaught of bourgeois ideas and the restoration of the bourgeois world outlook unless it stands on solid philosophical ground. In order to hold its own in this struggle and carry it to a victorious finish, the natural scientist must be a modern materialist, a conscious adherent of the materialism represented by Marx, i.e., he must be a dialectical materialist."[4] Lenin underlined that in materialist dialectics modern naturalists would find answers to the philosophical questions posed by the revolution in natural science.

Lenin interpreted the integrative processes in science not only as a purely theoretical, cognitive and methodological problem of systems in scientific knowledge, but also as that of the social mission of science and its interaction with the other social institutions and spheres of social life, a problem of the scientist's active social and ethical positions.

Through all of his theoretical and practical activities, Lenin assisted researchers in realizing the social significance of science so that they would consciously place its achievements at the service of the people, understand the laws of social development, and take an active part in building a socialist society.

Immediately after the 1917 Socialist Revolution, Lenin worked out the basic principles of the policy to be pursued by the Communist Party and the Soviet State with regard to science and formulated methodological guidelines for the correct choice of the main areas of research to solve the problem of planning and managing scientific activities. Lenin's Draft Plan of Scientific and Technical Work was in fact the first programme in the USSR and in the rest of the world for that matter for a planned and comprehensive development of scientific studies. It became a factor stimulating not only the development of the interaction between science and production but also the drawing together of various spheres of scientific knowledge. The Draft Plan dealt with the whole gamut of problems of the natural, technical and social sciences. Each of those problems outlined by Lenin was posed as a complex task which could only be dealt with by pooling different branches of knowledge. For example, electrification, the central problem in Lenin's programme of scientific and technical work, called for the combined effort of many different research teams. It exerted a noticeable influence on the development of many branches of technology, physics, chemical technology, applied mechanics, geography, and geology. The acceleration of the electrification effort involved the formulation and solution of a series of important theoretical and practical problems of economics, sociology, economic zoning, the cultural advancement of the outlying areas inhabited by the non-Russian nationalities, and profound changes in the life of the rural population, to mention but a few. The problem of studying the country's productive forces and their all-round utilization, closely linked with electrification, was of the same integral character. Significantly, it was when Lenin's plan of scientific and technical work was being carried out that comprehensive methods of organizing and conducting research became more and more widespread. The programmes for the study of the area of the Kursk magnetic anomaly, the Bay of Kara Bogaz Gol, the productive forces of the North, and others, were comprehensive in their content. They were oriented not only to the study of natural resources in those areas but also to examining their economic specialization and prospects for development, the ethnic and social composition of the population, cultural and historical traditions.

The national plan for the electrification of Russia known as GOELRO became a milestone in the developing interconnection between the natural, technical and social sciences. The drafting of the plan involved economic experts, engineers and naturalists, the best of the scientific community at the time. While the plan was being prepared under Lenin's guidance, research was carried out into the tendencies of progress in science, technology and production; the possibilities were assessed and the prospects were mapped out for developing large economic areas and the country as a whole. While the GOELRO Commission was at work, Marxist-Leninist theory began to permeate the natural scientific and technological studies of large research teams for the first time, harmoniously blending with their work. It is also significant that in the same years Lenin worked out and put into effect measures to promote the development of Marxist social science to meet vital practical needs. The idea was to create an integral complex of science in the Soviet Union.

It was on Lenin's initiative that the Socialist Academy of Social Sciences was set up in 1918. It was called upon "to elaborate the problems of socialism and communism and to conduct research in the spheres of social sciences, philosophy, and the sciences of nature, since the latter come into contact with the sciences of society".[5] The Socialist Academy was to translate into reality the idea of the synthesis of sciences in elaborating the practical and theoretical problems of socialist construction. A complex of scientific institutions took shape within the Academy's framework. Their purpose was to elaborate, from a Marxist standpoint, questions of history, the theory of state and law, the political economy of socialism, the problems of higher nervous activity, psychology, philosophy, and the methodology of science. Besides Marxist sociologists, taking part in the work of the Socialist Academy were such prominent natural scientists as O. Yu. Schmidt, A. N. Bach, S. I. Vavilov, and V. G. Fesenkov, to mention only a few. The centres of social sciences were brought under the authority of the USSR Academy of Sciences in 1936. The Academy formed the basis for the development of the countrywide system of fundamental scientific knowledge. So, the unity, co-operation and mutual enrichment of various branches of science were also cemented organizationally in the course of socialist construction.

The tendency of the various branches of scientific knowledge to become better integrated, for broader and closer co-operation among the natural, technical and social disciplines to channel comprehensive research into the mainstream of development in modern science, was theoretically substantiated in the works of the founders of Marxism-Leninism and already put into practice in the early stages of building a new society. At the present stage in communist construction this tendency is acquiring new traits and forms of expression in the conditions of developed socialism and on the basis of the unfolding scientific and technological revolution. The synthesis of scientific cognition in our time not only covers the related branches of science but also remote ones. This makes it possible to talk about the integration of scientific knowledge as one of the determining factors in its development at present-day level. Particular attention was paid to this circumstance in the decisions of the 24th and 25th Congresses of the CPSU, the need to strengthen the interconnection between the social, natural and technical sciences was underlined.

New forms of interaction between the theoretical and experimental means and methods of the social, natural and technical sciences are emerging in the course of the scientific and technological revolution. A case in point is the wide use of computer and information technology in the social sciences. At the same time, the reverse influence of the social sciences on the development of the natural and technical ones is increasingly making itself felt. Significantly, it is said more and more often now that scientific knowledge is becoming more humanitarian and that the humanities and social sciences are assuming ever greater importance for natural science and technology. For example, cybernetics has not only made a substantial contribution to the development of the methods of cognition in a number of social sciences. It has also enriched itself in interaction with them. Cybernetics experienced the influence of the general ideas of management and systems analysis advanced by progressive social and philosophical thought in the very course of its emergence. The notions of memory, training, games, collective behaviour and others entered cybernetics together with the themes and methods of psychology, sociology and linguistics. The ideas of the humanities are being introduced into cybernetics with particular intensity in view of the elaboration of the problems of "artificial intelligence", heuristics, and "communication" between man and machine. Such pan-scientific concepts as the study of operations, the information theory, the theory of systems, semiotics and the like serve as a powerful means of encouraging integration and strengthening the interconnections between the social, natural and technical sciences. The methods of these pan-scientific concepts are used in many spheres of knowledge.

In the course of the scientific and technological revolution the social sciences are becoming more and more closely and organically connected with production. This development opens up fresh opportunities for solving the important theoretical and practical problems linked with raising the efficiency of social production. Sociologists are now taking a very direct and active part in working out a theoretical basis for the scientific management of social and economic processes and studying factors helping to enhance creative activity, improve the organization of labour and production, and improve the system of material and moral incentives for the workers.

The specific features of modern technology, first and foremost, of the sophisticated man-machine systems, call for the use of the achievements not only of the natural and technical but also the social sciences in the very processes of their design and operation. The introduction of automation in production on the basis of computer technology substantially transforms the content and nature of labour. It requires the perception and analysis of vast amounts of information, complicated thinking operations linked with the work process, and implies the ability to make quick decisions in volatile situations. All this makes it necessary to study not only the physical chemical and technical properties of the working implements and materials to be processed but also problems related to the sphere of the social sciences. Scientists are facing the task of studying conditions ensuring the optimal co-ordination of man's physical, mental and intellectual potentialities, his aesthetic tastes and other social qualities with the demands presented by the modern and future technological systems. New scientific disciplines emerging on this basis such as ergonomics, technical aesthetics, applied linguistics, engineering psychology, and economic cybernetics integrate the theoretical and experimental means and methods of the social, natural and technical sciences.

In a developed socialist society the pooling of efforts by social scientists, naturalists, and experts in the technical sciences is first of all dictated by the need to develop all spheres of human activity in a comprehensive and planned manner and to manage the economy and public affairs in an efficient and scientific way. This interconnection is a must in solving such urgent problems as the substantiation of a long-term programme of scientific and technological progress and its social results; further elaboration of the problems of the USSR's fuel and energy complex, the optimal distribution of the country's productive forces, setting up large industrial and agrarian-industrial complexes of the type of the Baikal—Amur Railway, the West Siberian and the Angara—Yenisei projects. The idea is that of ensuring the maximum economy of labour resources, the supply of energy and raw materials, to improve living and working conditions, and also the rational use of natural resources, the protection and improvement of the environment in the interests of society.

In this context, the task facing Soviet scientists is that of working out a comprehensive approach implying the study of both man's influence on natural conditions and the effect of the environment on his health; the study of the structure and dynamics of the sum total of the biosphere's chemical, physical, and biological factors in their close connection with climatic conditions. An objective assessment of both the positive effect of the use of resources and the economic, social and even aesthetic damage it is causing must be made. Socialist society, relying on science and having placed natural resources at the service of man, is capable of solving the problems of the rational use of nature on a long-term, planned and genuinely scientific basis.

In the conditions of socialism the penetration of the social sciences into the sphere of production is not only linked with the need to raise its efficiency and to use new technology more rationally. The idea is also to create the best possible conditions for the all-round development of the individual, to free him from hard, monotonous work and to enable him to give full play to his creative potentialities in his labour activity.

The study of present-day global problems constitutes an important sphere of interaction between the social, natural and technical sciences as set out in the decisions of the 25th Congress of the CPSU. These include the problems of food, energy and raw materials, stamping out the most dangerous and widespread diseases, space exploration, and the use of the world's oceanic resources. In every sphere of knowledge, Soviet scientists are called upon to make a weighty contribution to solving these problems affecting the interests of all mankind.

Special attention is now being paid to the social and economic assessment of various achievements in techniques and technology, the elaboration of scientific criteria of the efficacy of quality, the optimal correlation between the long-term programmes and the solution of the problems capable of yielding quick and substantial results.

Experience shows that the elaboration of long-term scientific technological, social, and economic prognoses is a highly effective form of teamwork by experts in different sciences. Significantly, the 25th Congress of the CPSU paid much attention to drawing up and implementing comprehensive programmes. One such programme of scientific and technological progress, taking into account its social and economic consequences, is intended for the period up to 1990 (and even up to a later date in some cases). It is called upon to provide the quidelines indispensable for the successful planning of the country's social and economic development. The main directions of scientific and technological progress outline in the programme are to become a basis for co-ordinating research for many years to come.

In the light of the decisions adopted by the 25th Congress of the CPSU the entire organization of research is to be raised to a qualitatively new, much higher level to enable the country to move to the fore in the key areas of world scientific and technological progress. This also applies to the organization of efficient co-ordination between the natural, social and technical sciences. The work of the scientific co-ordinating councils must be raised to a new level, and they must be provided with the appropriate facilities. Evidently, in view of the tendencies in the development of contemporary scientific knowledge, it will be vital to practise more widely the setting-up of scientific teams and branches made up of specialists in various fields of the natural, technical and social sciences to solve one complex scientific or technological problem or another. Such fundamentally flexible outfits would be free from the shortcomings of the rigid structure adopted in branch research centres, a structure that sometimes remains unchanged for decades.

The organizational structure of science in the Soviet Union is based on specialization according to subjects and is therefore of a "branch" nature. This structure is quite justified and it remains the principal form of the concentration and co-ordination of research. Yet it is apparent that the development of social practice and cognition is posing more and more problems; the solution to these problems requires going beyond the boundaries of the individual specialized scientific institutions. As a result, it is becoming necessary to organize research in a comprehensive manner, according to the problems tackled, which is typical of our time.

However, extreme views are already making themselves felt in this field. It is alleged that the boundaries between the sciences should be effaced and that it is obsolete practice to divide the sciences according to the subject of their study.

Marx's aphorism about one science in the future is being interpreted simplistically in this context. However, the entire theoretical work and research done by Marx, Engels and Lenin indicate that what they meant was the common methodological bases of scientific cognition and the inevitability of their increasing organic synthesis, and not the replacement of all sciences by one. Present-day science is moving precisely in this direction.

Firstly, we cannot fail to see that besides the integration of human knowledge the differentiation of sciences is continuing, which reveals the dialectical nature of scientific cognition. Secondly, it is wrong to identify a science with its object of study. It is common knowledge that many sciences have common objects of study such as nature, society, man, coupled with more concrete subdivisions into inorganic or organic nature, plant or animal life, etc.

Each science has a subject of its own, i.e., it studies specific law-governed patterns operating in the given field or in the entire objective world, such as physical or chemical patterns, etc. Suffice it to recall the negative influence exerted on the study of human consciousness by the identification of higher nervous activity and psychology linked with the erroneous interpretation of the results produced by the session of the three Academies devoted to Pavlov's teachings. The view prevailed at that time that psychology had no subject of its own since it studied the same nervous activity as Pavlov's physiology. In fact, however, they are two different sciences, although they do have a common object of study. One of them studies physiological processes while the other deals with mental processes and the law-governed patterns of higher nervous activity. The successful development of psychology was made possible only after the special subject and individual status of that science, with the specific law-governed patterns of its own, had been recognized. However, its link with the theory of higher nervous activity as an outstanding scientific achievement is becoming firmer rather than weaker. This is one of the many cases in point testifying to the dialectical process of the differentiation and integration of knowledge.

At the same time, as far as genuinely comprehensive scientific research is concerned, it would be an obvious oversimplification to reduce the whole thing to the study of one object or another by the methods of different sciences and the subsequent summing-up of the results. To summarize what has been done by many sciences in studying the same object is an important, yet only an initial stage in comprehensive research followed by another, which is more difficult and complex. This is to create a whole theoretical picture of the phenomenon under study and work out concrete scientific recommendations for practical purposes on this basis. In other words, the need arises to develop a specific conceptual machinery, the logic and methods of comprehensive research, as it were, which would act as a means of organizing research, a means of "bringing" various "individual" models of the object into its full, integral picture. This function is assumed by the interdisciplinary sciences that have emerged in recent decades such as cybernetics, semiotics, the information theory, etc. Without attempting to analyse these logical and methodological means, let us underline just one circumstance. These latest methodological means of interdisciplinary synthesis are profoundly interconnected with dialectics. This interconnection goes back to the origins of the systems approach and other fruitful methodological trends of our time. Contrary to some of the unjustified claims and misgivings, neither the systems approach, the information theory, nor semiotics can ever compete with materialistic dialectics, much less so, replace it. What is more, it is in the ideas of dialectics that they find their substantiation.

The interconnection between philosophy and concrete sciences and practical thought is becoming ever more sophisticated. Philosophy is developing as an organic element of the integral system of the knowledge of the world. Its importance as a methodological basis for integrative processes in the sciences is steadily growing. All this constitutes one of the most vital aspects of the integration of contemporary scientific knowledge.

It must be pointed out here that Marxist Leninist philosophy occupies a special place among the philosophical systems and concepts. It is the only philosophy for which integration processes are the natural and logical manifestation of its own inner life. The reason is that the very possibility of organically including philosophy in the integral system of the scientific knowledge of the world arises in so far as philosophy recognizes the scientific nature of its teaching and consistently brings it to life. On the other hand, any negation of the scientific nature of philosophy automatically excludes philosophical thought from the general process of the integration of knowledge.

It is an open secret that many bourgeois philosophers deny the scientific nature of philosophy and set philosophical thinking against scientific thought and cognition. This drive to separate philosophy from science first of all stems from attempts to strip the latter (social science in the first place) from any elements of ideology or Weltanschauung and, consequently, exclude it from the context of the present-day social struggle. Indeed, if we recognize philosophy as a science, the invevitable conclusion is that, on the one hand, it is possible to achieve the unity of the methodological principles in the various types of man's cognitive attitude towards the world and, on the other, to create the only true picture of the world from a philosophical point of view. This reveals the erroneous nature of cognitive and practical relativism, the fundamental pluralism of the vision of the world that constitutes one of the basic dogmas of subjectivism and idealism in contemporary bourgeois philosophy.

While acknowledging the difference between philosophy and the specialized sciences we proceed from the fact that materialistic dialectics is a science.

The ideological and world outlook function of dialectical materialistic philosophy is far from ruling out its scientific significance. This function is not something external, inorganic with regard to materialistic dialectics as a scientific philosophy. On the contrary, it stems from this characteristic of dialectics and is determined by it. Moreover, this is what stipulates the organic link between philosophy and the sciences of nature and society.

The problem of the integration of philosophical knowledge into the general system of the scientific knowledge of the world manifests itself first of all as that of the functions of Marxist-Leninist philosophy with regard to science.

Firstly, philosophy poses as a science summing up knowledge and discovering the general principles which underlie the latter. The very possibility of philosophical synthesis is striking evidence of the fundamentally integral and systematic nature of man's knowledge of the world. Although this integrity cannot find its complete expression at every given moment on the level of the concrete sciences because human cognition is incomplete, it does find its substantiation in the very existence of the philosophical knowledge of the general.

Of course, it is the objective of any science to discover laws and that is why it is engaged in generalization. However, in a concrete science generalization is confined to the field it studies. For example, the notion of heredity is very important for the development of scientific cognition. Yet it holds good only for the organic world, making it possible to understand the evolution and activity of the living organisms and is not applicable to inorganic nature. Nor can it be applied to the development of human society, the origin and behaviour of class and social groups.

Among the specialized sciences it is mathematics that produces the most extensive generalizations. Its means and methods are applicable to any field — economics, astronomy, sociology, and biology. Mathematics plays a very important role not as a method of describing phenomena but also as a means of discovering new truths. In the most advanced areas of theoretical natural science, mathematics has become the principal form of expressing the appropriate law-governed patterns of nature. However, mathematical generalizations largely hold good for quantitative characteristics. Therefore, although they are effective in a very wide sphere and possess tremendous potentialities for analysing quantitative relations, they cannot be regarded as universal.

Logic with its numerous branches plays a major role in the generalization of scientific data. Logical terms and rules of thinking help in systematically analysing scientific problems and enable precise conclusions to be drawn.

However, a feature of philosophical generalizations (as compared with those made in the sphere of the concrete sciences) is that they are universal. At the same time, the most important task of philosophy is to elaborate scientific methods of generalization and also a scientific world-outlook and methodology.

It should be pointed out that the mathematical, logical and other methods of generalization cannot work out solutions for such general philosophical problems as those of being and thinking, the subject and object, man and nature, nature and society, theory and practice. The same is true of a series of other general problems of Weltanschauung and methodology, which are tackled precisely by Marxist-Leninist philosophy. Without their solution it would be impossible to use logic and mathematical methods in cognition. That is why the elaboration of the general philosophical categories and laws helps in understanding correctly and in developing the entire generalizing mechanism of present-day scientific cognition.

Secondly, philosophical generalizations are universal. They lie on a fundamentally different level unlike the positivist inventories of conclusions obtained "for the given moment" in the mass of the concrete sciences. As a result, these generalizations determine the scientific and heuristic role of philosophy. The latter does not claim it can solve the specific problems of the concrete sciences. Yet it is developing in close contact with them. In its notions and categories it sets out the logic of scientific cognition, which is a profoundly dialectical process by nature, and serves as the methodological basis of the Marxist concept of scientific development.

The positivists deny the possibility of obtaining new knowledge within the framework of the philosophical interpretation of the world. As a result, they actually reduce the problem of the integral picture of the world to a flat, single-level model in which generalization is interpreted as a purely quantitative expansion of the volume of notions used in formal logic.

Of course, there is a formal method of generalization when the individual, specific features of objects are brushed aside and ignored to obtain general names for the latter.

However, in Marxist-Leninist philosophy, generalization is understood as the cognition of ever deeper connections and phenomena and not as a purely "quantitative" process: generalization is a transition in thinking from one level of objective interdependence to another, more general and profound.

Works by the classics of Marxism-Leninism tell us that the philosophical generalization of data coming from science and social life helps elaborate and substantiate fundamentally new notions and ideas. In this connection, suffice it to recall the immense importance for scientific cognition of Marx's notion of the socio-economic system as a logical stage in the history of human society. The Marxist-Leninist concept of history based on this notion has formed a firm foundation for the scientific cognition of social phenomena. We might also recall that Lenin, summing up the latest achievements in physics in the early twentieth century, came to the conclusion that the electron was as inexhaustible as the atom. This generalization was an outstanding philosophical discovery confirmed by the whole subsequent development of natural science.

Last but not least, philosophy serves as a methodological and organizing centre of connection and mutual influence between different scientific disciplines. Indeed, a number of fundamental notions in the cognizance of the world are developing as a category of philosophy and the individual sciences, for example, matter, space, time, motion, causality, quantity, quality, etc. By elaborating the content of these categories, philosophy creates a mechanism of notions to form a vision of the world in various branches of knowledge. At the same time, philosophy also deals with the real problems originating in one individual branch of knowledge or another. Since philosophical thought is not handicapped by a narrow "professional" approach to those problems, it discovers their general importance, stimulates interest in them in other branches of scientific knowledge, helps develop interdisciplinary contacts, etc. This methodologically organizing role of philosophical thought is becoming more and more apparent today. Comprehensive research based on the interaction of the natural, technical and social sciences as well as philosophy is now well under way in studying the biosphere, social and economic development, and the problems of man. This interaction is vital for comprehensively solving the problems of scientific and technological progress, the development and rational distribution of the productive forces, the construction of large industrial and agrarian-industrial complexes, urban and rural development, creating optimal conditions for man's life and work, and environmental protection.

The comprehensive nature of problems facing scientific cognition in our time makes even more urgent the need to create an integral scientific picture of the world as mentioned by Lenin in his brilliant work Materialism and Empiriocriticism.

This need is not only felt within the framework of the philosophy of dialectical materialism. It also manifests itself in various trends of modern bourgeois philosophy in one way or another.

The neo-positivist programme of "integral science" is a case in point. Neo-positivism, being a variety of subjective idealism, sought the unity of scientific knowledge exclusively in the sphere of the cognizing subject, in his sensations and psycho-physiological organization, intellectual abilities, cognizing operations, language, etc. What was entirely ignored was the key role of the material unity of objective reality with regard to that of scientific knowledge.

However, this interpretation of the unity of science is self-contradictory. The contradictions became particularly pronounced in one of the positivist trends, which attained a definite influence in the first half of this century. That was the philosophy of physicalist analysis. Physicalism urged that metaphysics should be banished from science (in its positivist interpretation as everything which is not reduced to the data of direct observation) and the creation of a single language of science. According to the views expressed by the adherents of this conception (O. Neurath, R. Carnap, K. Hempel, etc.), any proposition can be regarded as scientifically sound only if a method has been demonstrated of reducing this proposition to that of the most elementary particles known at the given stage in the development of scientific cognition. Consequently, the physicalists maintained, the only way to synthesize our knowledge of the world is to translate all the scientific propositions into the language of the physics of elementary particles, which was considered the prototype of the future single language of science. Any attempt to study a sphere of reality on the assumption that it has specific qualitative features of its own was described by the physicalists as metaphysics.

Meanwhile, it is not hard to see that the programme in question was itself based on a metaphysical postulate. It is a postulate of extreme reductionism maintaining that all forms of the motion of matter can be reduced to those studied by physics, in particular, the physics of elementary particles. It is apparent that this interpretation of the unity of science is not only self-contradictory. It runs counter to the real tendencies in its development. Suffice it to imagine the sorry plight a biologist, a psychologist or a sociologist would find himself in if he tried to put this programme into practice in his research. He simply would not be able to formulate a single statement related to the objects he was studying. Small wonder, therefore, that even the founders of physicalism were later compelled to ease the rigidity of their original tenets gradually, although this did not save physicalism from its ultimate collapse.

So the integration of the sciences and the question of determining its most fruitful trends is far from being confined to stating individual facts of history and the present state of science, no matter how important these facts might be. The same also applies to the foundations of the interconnection between the social, natural and technical sicences. All these constitute a major philosophical and methodological problem.

When Marxist-Leninist philosophy underlines the unity of science, it means first of all the unity of the methodological foundations and not the elimination of the historically established spheres of knowledge. These foundations make it possible to overcome finally the age-old division of individual sciences and substantiate scientifically the principles of co-ordination. This unity of methodological foundations is provided by dialectical materialism, a scientific world outlook that is synthetic in its nature.

Two extremes manifest themselves in pre-Marxist and contemporary bourgeois philosophy in interpreting the interaction between philosophy and concrete sciences.

One of them attributes to philosophy the ability of the monopoly cognition of the nature of things and the world as a whole. Yet in this case there is no place for other sciences. There was a time when philosophy included all the spheres of knowledge or, to be more exact, the embryos of knowledge in the fields of nature and social life. Yet specialized sciences branched off from philosophy in the course of history. The natural sciences study various spheres of nature. Social life is studied by a whole set of economic, social political, and humanitarian sciences. Philosophy never replaces all the spheres of knowledge. Nor can it do so. The time of natural philosophy and the comprehensive philosophy of history passed long ago.

The reaction of nineteenth and twentieth century bourgeois philosophy to the natural philosophical traditions and scholastic metaphysical conjectures of religious philosophy gave rise to the other extreme. The latter was most manifestly expressed in the emergence and development of various positivist schools and trends. This extreme is in fact reduced to a denial of the scientific nature of philosophical knowledge, of the close, unbreakable link between philosophy and the rest of the forms of scientific cognition of objective reality. According to the positivist interpretation, philosophy, if it has any claims to a scientific nature, should cast aside problems of Weltanschauung and remain just a specific form of concrete scientific knowledge.

While the supporters of the natural philosophical trend or religious philosophical metaphysics claimed a *Weltanschauung* synthesis of a sort, the adherents of positivist philosophy retired to individual, highly specific areas, having brushed aside the general world-outlook contents of philosophy. If they ever attempt to evolve a general method, they confine themselves to the specific problems of phenomenology, structuralistics, logical-mathematical, "linguistic" analyses, etc. As a result, positivism has failed to form an integral system of Weltanschauung or a consistent theoretical and methodological basis for scientific cognition. It is modern positivism in its different forms (logical positivism, operationalism, pragmatism, etc.) that has been most instrumental in pushing bourgeois philosophy away from problems of world outlook. It has contributed to the formalization of the epistemological problems and the development of the "technique" of philosophical analysis to the detriment of the solution of the fundamental methodological problems of cognition.

Marxist-Leninist theory rejects the natural philosophical and positivist concepts of the interaction between philosophical and concrete scientific knowledge. As a result, it overcomes the metaphysical opposition of philosophy and individual sciences as well as the gap between the methodological an world-outlook functions of philosophical science. Dialectical materialism views the subject of philosophy not as something isolated from the concrete forms of cognition but as its result stemming from the theoretical understanding of objective reality by the subject. Dialectical materialistic philosophy performs its Weltanschauung function by generalizing the conclusions of concrete scientific cognition and discovering the general in the specific.

The Weltanschauung function of philosophy is inseparably linked with the methodological one. Marxist philosophers proceed from Lenin's idea of the fundamental unity of the characteristics of being, objective reality, and of the cognition of this reality. This means that, while studying objective reality reproduced by science, i.e., analysing the content of scientific discoveries, philosophy is also studying the forms of their theoretical expression, the methodology and logical structure of research. The general laws and categories of materialistic dialectics, logically expressed in the form of a philosophical theory, assume the nature of methodological prerequisites with regard to scientific theory to become the principles of scientific cognition. So by evolving a scientific view of the world materialistic dialectics also becomes a means of attaining new knowledge.

To be more specific, Marxist-Leninist theory singles out two dialectically connected streams flowing in different directions in the interaction of philosophy and individual sciences: from particular sciences — through the generalization of their theoretical conclusions — to philosophy, a scientific theoretical world-outlook, and from philosophy — by evolving the methodology and logic of research — to particular scientific cognition.

Scientific knowledge undergoes radical changes at a time of revolutionary development. New theoretical explanations of the phenomena and links of the objective world appear by leaps and bounds. When they are summed up in the theory of materialistic dialectics, a system of scientific world-view, they complement and deepen man's most general views of objective reality. At present, Marxist-Leninist philosophy is facing the theoretical task to generalize the current scientific and technological revolution. This revolution is radically changing not only the entire fabric of theoretical and applied knowledge, but also the place of science in the structure of mankind's productive forces.

Scientific revolutions are most productive as far as the exercise of the *Weltanschauung* function of science is concerned. Yet they prove to be most difficult in exercising the methodological function of philosophy. The point is that its methodology can lead individual sciences successfully only if it can catch up with the new theoretical notions and steadily get rid of the obsolete "structures" rejected by the new theoretical conclusions of the individual sciences.

In view of the universal nature of its premises and principles, Marxist-Leninist philosophy is capable of playing this methodological role in the above-mentioned conditions. As a special type of theory developing a logically substantiated system of categories and law-governed patterns, reflecting some of the general principles of being and consciousness as well as relations between them, materialistic dialectics becomes an important factor in theoretical elaboration in every sphere of modern natural scientific and social cognition.

There are different levels of research in the present-day methodology of science. Each of them has its own ways and means of analysing the cognitive process. Far from all of them can and should fall within the category of the philosophical ones. A considerable part of the methodological elaborations carried out by non-philosophical ways and means (e.g., with the aid of the methods of formal logic) possesses every essential trait of concrete scientific knowledge and cannot claim the role of the general methodology of science. It is only philosophy that can assume this function. It is materialistic dialectics that is exercising it in the most effective manner in present-day conditions. In studying nature and the possibilities of scientific cognition as well as the relationship between scientific knowledge and reality, Marxist-Leninist philosophy is searching for the general foundations of science, the conditions of the truthfulness and authenticity of scientific knowledge, its prerequisites, and the mechanisms of its formation. Therefore it objectively acts as the general methodology of scientific cognition.

Besides this, there are also specialized ways and means of conducting research used in concrete sciences. They exercise the function of additional methods regulating the advance of cognition to the truth. The very possibility of using different methods in studying man's cognitive activity as well as the evaluation of the results rely on definite philosophical prerequisites and notions, which are those of general methodology in essence. Therefore the development of the specialized scientific means of analysing methodology is far from depriving philosophy of its role as a general methodological basis. Nor is it just turning it into a branch of specialized scientific knowledge. Any attempt to build a more or less integral system of methodology cannot but lead to an analysis of such general problems as the nature of knowledge, the character of its relation to reality, the nature of truth and ways of verifying it, etc., that is, the very same problems that are tackled by philosophy. The collapse of the logical positivist concepts of science vividly shows that it is impossible to build a general methodology of science as a specialized scientific discipline divorced from or even opposed to the philosophical analysis of cognition.

The dialectical connection between philosophy and the concrete sciences consists in the fact that as philosophy is being enriched and perfected through the generalization of data and theoretical conclusions produced by individual sciences, the latter, for their part, are able to move ahead with a greater measure of success due to the development and improvement of dialectical materialist theory. Therefore one of the crucial tasks for Marxist philosophers at every stage in the development of cognition has been and remains the advancement of the theory of materialistic dialectics, the philosophical interpretation from the point of view of *Weltanschauung* of the conclusions and achievements of the individual sciences, the creative elaboration of the methodology of scientific cognition and the revolutionary transformation of reality.

REFERENCES

1 K. Marx—F. Engels, *From Early Works*, p. 596.
2 K. Marx—F. Engels, *Works*, vol. 20, p. 43.
3 V. I. Lenin, *Collected Works*, vol. 20, p. 196.
4 V. I. Lenin, *Coll. Works*, vol. 33, p. 233.
5 *Soviet Government Decrees*, vol. II, Moscow, 1959, p. 469.

2. Philosophy and the Synthesis of Cognition

HERBERT HÖRZ

Philosophy has always been concerned with the synthesis of knowledge. As long as sciences with special objects of cognition were lacking, philosophy filled in the gaps in knowledge by speculation. The tempestuous development of sciences, their penetration into all spheres of natural and social existence makes the task of philosophical synthesis obvious. Philosophy can be no substitute for science, but it must rather answer by scientific methods the fundamental questions in world-outlook. World-outlook is a system of answers to questions about the origin and nature of existence of the world, about the source of knowledge, about the position of Man in the world, about the sense of life and the character of social progress. Philosophy generalizes practical experience and social knowledge in order to answer these questions. Philosophy is the theoretical basis of world-outlook and must analyse cognition, in order to be able to synthesize the analysed essential elements.

2.1 TASKS OF PHILOSOPHY

With the introduction of the scientific and technological revolution, science pervades more and more all spheres of life. General theories of vast scope are formed. Complex systems are explored theoretically. In the focus of the synthesis of cognition is Man as a subject and object of history, as a recognizing, evaluating and active subject. Natural, technical and social sciences supply material to philosophy for the synthesis of cognition. Philosophy itself has an impact on world-outlook, science and practice by scientific insights. In this context, philosophy has to answer the old questions: What must we know? What can we hope for? What shall we do? What is Man? The answers to this have acquired new aspects due to the tempestuous development of the sciences, and these aspects must be investigated. Science is individual and organized social activity to recognize the objective relations and laws of nature, of society and of consciousness in their systematic and historical interrelation and in their significance for social and personal development (science as a cultural force). Science utilizes these insights for the production of new substances, processes and machines, so as to make more effective, directly and indirectly, the production of material goods (science as a productive force) science analyses the structure of social relations necessary for social progress, and science creates certain prerequisites for a conscious humane orientation of scientific and technological advance (science as a social or humane force). In this context, there emerge philosophically relevant tendencies for the production of cognition. These include:

Firstly, the social relevance of results obtained in the development of science is increasing. Scientific and technological advance creates prerequisites for social advance. Thus the relation between humanism and the development of science becomes a focal problem of discussion on world-outlook. Humanism becomes a target function in socialism, an evaluation criterion and a strategy of demands for scientific and technical advance. Philosophy is called upon to set free ideal driving forces for a conscious humane orientation of scientific and technical advance and to eliminate obstacles in world-outlook.

Secondly, the further development of socialist society is necessarily linked with an increase in performance of the socialist national economy. This requires original research programs which lead to new insights and raise the prestige of socialist science, and this demands principal solutions with a high social effectiveness. Marxist-Leninist philosophy must prove a reservoir of creative ideas by promoting with its own means the development of science as a productive, a humane and a cultural force.

Thirdly, the solution of complex scientific problems connected with complexes of social requirements, such as the environment, energy, raw materials, health, nutrition and personal development, calls for co-operation between natural, technological and social scientists. It is necessary to investigate the interplay of natural, social, technological and cultural factors. Marxist-Leninist philosophy provides an ideological, epistemological and methodological — including a terminological — basis for this interdisciplinary and interinstitutional co-operation.

The synthesis function of philosophy therefore has various directions. This function is firstly aimed at *world-outlook* and hence at understanding the human being in his position within the world. Knowledge must be synthesized in such a way as to increase the cognition of Man by himself as the subject of history. Secondly, this function is aimed at the development of the ***sciences.*** This function must by its specified philosophical statements and philosophical hypotheses stimulate the advance of sciences, hence, acquire a heuristic effect. Philosophy can only do this by analyzing and generalizing the most recent insights of sciences in philosophical terms. Thirdly, philosophy as a decision aid in terms of world-outlook aims at *social* practice with its synthesis of cognition. Philosophy acts in the formation of motives and desires. Human beings must make their proofs in their own activity by competent decisions as free subjects. Directly linked with these developments of synthesis are the functions of Marxist-Leninist philosophy. Philosophy must examine its role towards the development of science if it wants to do justice to the above-mentioned tasks. It is necessary to synthesize insights in such a manner that they can become scientific foundations for well-founded answers to basic questions of world-outlook in our time (world-outlook function of philosophy). The connecting links between general philosophical statements and some certain special scientific insights must be utilized in such a manner as to promote the development of scientists by means of heuristic suggestions (heuristic function of philosophy). From philosophical insights it is necessary to obtain substantiated scientific orientations for action which as mental driving forces have an impact on action and thus have ideological repercussions (ideological function of philosophy).[1]

As there are no logical derivations for philosophical insights from insights of special scientific disciplines and no algorithm either which transforms philosophical statements into instructions for action, the process of philosophical generalization must be analyzed in its various components. Only in this way can the complicated scientific cognition process of Marxist-Leninist philosophy be understood, which must be aware both of the danger implying a depletion of content for its categories and a neglect for the synthesis of knowledge.

2.2 THE PROCESS OF PHILOSOPHICAL COGNITION

Marxist-Leninist philosophy gives substantiated scientific answers to basic questions of world-outlook by generalizing scientific insights from the development of the revolutionary class struggle and from the cognition of nature and society. Thus the unity of the three functions becomes apparent. A scientific world-outlook which realizes its mobilizing effect on the masses, hence the ideological function from the objective truth of its fundamental statements is based on a scientific philosophy. Scientific insights are investigated and therefore their philosophical analysis can assume a heuristic significance for scientific work itself.

As some of the basic principles which distinguish Marxist-Leninist philosophy from all other philosophies we understand the following: Philosophical materialism with its recognition of the primacy of matter over consciousness, the material unity of the world and the existence of objective laws in nature and society. These include materialist dialectics recognizing the objective interconnection, motion and development which takes place on the basis of dialectical opposites as a transition from one quality to another by quantitative and qualitative changes within the old quality leading to the dialectical negation of the negation. They also include historical materialism which investigates the conditions of production as determining social conditions and explores the dialectics between the basis and the superstructure; historical materialism explains the history of society as a history of class struggles which have proved to be the driving force of social development ever since the creation of private property and the state and which will be abolished only by the construction of a classless society. These principles embodied in the works of the classics are the startingpoints of the philosophical analysis and synthesis of scientific insights and they have to make their proofs in this. The first step of philosophical work is the defence of these basic principles by proving the compatibility of philosophical statements with scientific insights. Because the basic principles are highly general, mostly no great expenditure is necessary for the most part for this purpose, although in discussing world-outlook this proof is significant for rejecting vulgar, one-sided and distorted views on our philosophy. We have only to remember discussion, say, about the theory of relativity in which the relativity of space and time related to individual motion, hence the denial of absolute space and of absolute time was intended to be utilized against materialism. Unfortunately, a few Marxists rejected the theory of relativity with this argument because they failed to analyse the incorrect philosophical arguments objectively. However, the thesis about space and time as a form of the existence of matter requires a precise investigation of the objective space-time structure dependent on the motion of material objects and expresses the basic materialistic view that our space-time-theories are a reflection of objective space-time relations. This proof of compatibility as an expression of the synthesis of cognition is the prerequisite for a further philosophical analysis of the space-time relations leading to views about the relation between matter, motion, space and time and making it possible to recognize space more and more as a real objective structure. Insight into relations connected with the discontinuity and continuity of space-time, the question of spatial divisibility, the irreversibility of time, etc. are also being investigated. Through philosophical statements presented in more concrete terms we understand the basic principles that are extended and explained by means of the accumulated knowledge of an era.

In order to be fruitful for science, each general philosophical statement must be presented in greater detail for that sphere in which it is to be utilized. This is only possible by analyzing the scientific material from this sphere. For instance, let us consider the relation between law and chance in its significance for physics. By law we understand a generally necessary, i.e. reproduceable and essential interrelation which determines the character of a phenomenon. This definition can be presented in greater detail for classical physics to the effect that a law is a possible relation which must necessarily be implemented by the physical process in conditions where the law applies. Then, chance will be a concrete manifestation of this necessity in which variations occur arround the tendency of behavior, expressed in the law. This view of laws, which has been presented in more precise forms for classical physics, is no longer sufficient for the insights of quantum mechanics. New natural science material demands a new, more detailed definition of the philosophical concepts. Approximately the following more detailed definition would do justice to the quantum theory: A statistical law is the existence of a system's possibility which is necessarily put into effect, the existence of a series of mutually connected relations of which one relation is accidentally put into effect by an individual object, whereas the entirety of the objects belonging to the system necessarily puts into effect this series of possibilities. The law indicates a probability for the accidental implementation of a possibility.

The views on laws put into more precise terms for classical physics and for quantum physics represent two types of a law formulated here in such a manner that they are also applicable to other processes. In physics, the classical (dynamic) law proves to be a limit case of the statistical law. Thus also in statistical law the dynamic aspect is contained, owing to the existence of the system's possibility which is necessarily put into effect. In this manner, it could be possible to define more exactly the view about a statistical law: A law is a generally necessary and essential interrelation (fundamental principle) which indicates such a possibility of the system's behaviour necessarily put into effect (dynamic aspect). By contrast, for the behaviour of the elements there exist various possibilities with a probability distribution (stochastic aspect), and each accidental of a possibility takes place with a certain probability (probabilistic aspect). In this manner, the philosophical concept of the statistic law has been defined. This concept synthesizes all insights about laws. For further research it is necessary to construct philosophical hypotheses. A philosophical hypothesis is an assumption formed in scientific terms from the results of science and from their philosophical analysis on the validity of interrelations of one science in another or on the contribution of the developing theory to a more exact definition of philosophical categories. In order to state the general validity of the above-mentioned types of laws, it would be completely incorrect to consider them as already generalized after an analysis of physical theories. A necessary degree of generalization would be the hypothesis that these types of laws also occur in biology or in social scence. Each of these hypotheses must be examined in detail. The hypothesis, however, can also refer to a more exact future definition of philosophical concepts by science. It would also be possible to put forth hypotheses about the relation of these two types of laws in future physics. From these few remarks the complicated nature of a philosophical generalization becomes obvious. Apart from defining philosophical concepts more exactly by means of confirmed philosophical theories and from hypothesis formation based upon this, it is also necessary to analyze the physical hypotheses according to the philosophical view of laws which is connected with this.

The investigation of statistical laws leads to a new relation between law and necessity. The law also comprises possibilities that are accidentally put into effect for whose implementation of the statistical laws indicate probabilities. The entirety of relations necessarily put into effect by all elements of the system to which the statistical law applies as a system law and the possibility that is brought about accidentally by one element proves the dialectics of necessity and chance in the law.

The philosophical generalization of scientific insights cannot take place today as in the times of Marx and Engels. Scientific material is no longer as easy to survey as in the past century. Conceptions have failed that foresaw a general theory of the world, of society and of nature.

The philosophical generalization of scientific insights is carried out in a unity of deduction and induction, of analysis and synthesis, by defining more exactly philosophical statements for one field of research through theories existing in this field. This provides a basis for a set of philosophical hypotheses about the development of science and about the validity of a more exact definition in other fields as well. Only after confirmation of these hypotheses is the universal validity of the more exact definition proved, and it is then included in the inventory of Marxist philosophy, an inventory which is the basis for further research.

Hence, Marxist philosophy consists of statements with a varying degree of generality. Its basic inventory is relations and laws with a significance in terms of world-outlook. The basic questions in world-outlook are answered by Marxist philosophy via such partial theories as the theory of the material unity of the world and the theory of dialectical materialism, the theory of development, the theory of society, etc. Therefore, the generalization of insights in natural science must be carried on until the decision is reached as to whether they represent new parts of the basic inventory of Marxist philosophy. This basic inventory was elaborated by Marx, Engels and Lenin, and it provides the foundation for the philosophical analysis and synthesis of the scientific material. However, general statements must be defined more exactly.

Apart from the basic inventory there exist philosophical statements for certain fields of research and these statements were defined more exactly by means of insights and discoveries of sciences in an era. It is necessary, here, to examine the statements substantiated earlier and linked to the results of science of the last century.

Statements that are defined more exactly are the basis for philosophical hypotheses, referring either to the development of the field of research, to which the more exactly defined statement applies, or aimed at universal validity of the recognized state.

More precisely defined philosophical statements and philosophical insights are a necessary, integral part of philosophical cognition, whether aimed at enlarging our basic inventory of philosophical statements or at assisting the development of sciences. Hence the synthesis of scientific insights is a necessary prerequisite for fulfilling the function of philosophy.

2.3 FUNDAMENTAL PRINCIPLES OF MARXIST-LENINIST PHILOSOPHY AS A SYNTHESIS OF KNOWLEDGE

The fundamental principles of Marxist-Leninist philosophy have been elaborated in a long historical process of development in science. On the one hand, they are a synthesis of previous knowledge, and on the other, heuristic processes for obtaining new knowledge. What are such fundamental principles? Dialectical materialism is the unity of philosophical materialism and of materialist dialectics. Materialism means to recognize the facts in their own and not in a fantastic interrelation; hence, to disclose the form of the objective interconnection and the emergence of the new in qualitative terms. Consistent materialism implies dialectics. Dialectics comprises the material unity of the world not in such an intertwined chaos of relations in which the cognition of a structural element, be it a solid or an organism, can only be obtained if the cosmos were cognized. In this meaning, however, it would be possible to understand the thesis that all is interconnected with everything. But cognition about the objective interconnection of objects and processes says that there is no material sphere which is not linked with other spheres by material processes. Compatible with this is also the existence of a finite velocity of signals and effects. The insights of dialectical materialistic monism can be summarized in fundamental principles, that is in general principles of theoretical and methodical behaviour which are based on objective dialectics:
(1) the principle of the inexhaustibility of matter;
(2) the principle of the structuring of matter;
(3) the principle of dialectical determinism;
(4) the principle of development.

These principles are further specified by the development of science, whereby materialist dialectics absorbs the insights of pre-Marxist philosophy. The insights of materialist dialectics are a unity between a descriptive picture and creative construction. The last criterion of the truth of dialectics is social practice. Dialectical relations and laws cannot be proved or refuted in their generality unrelated to science and reality, but they can be proved or refuted in their concretized form.

2.3.1 Principle of the Inexhaustibility of Matter

Until the 19th century, the infinity of matter was essentially understood as being boundless. This argumentation was used by Engels in a justified manner against Dühring's affirmation that the world had a beginning in time and was enclosed in spatial limits.[2] But Engels also referred to Hegel's criticism of bad infinity pointing to cognition about the infinity of a finite number of objects in the form of the universality of objective laws. Because motion is the manner of existence of matter, an eternal change of form of matter takes place. Objective reality is inexhaustible in its objects, processes and relations.[3] The exactly defined philosophical statement about infinity as being boundless had already proved problematic with the work of Lobachevsky, Gauss, Riemann and Helmoltz about non-Euclidean geometry, because the boundlessness of finite surfaces could be shown. But even dialectically trained thinkers did not always see the significance of this work. Thus, N. G. Chernyshevsky wrote in 1878 about the work of Helmholtz "On the origin and importance of geometric axioms"[4] that it contained "senseless stupid things" because it was claimed that new systems of geometry could be created. This was considered incompatible from the materialistic viewpoint, and, therefore, Chernyshevsky emphasized: "This is the wild speculation of an *ignoramus* who does not know what he thinks and what he thinks about."[5] With Einstein's General Theory of Relativity, the physical relevance of non-Euclidean geometry became obvious. This led to many discussions about the philosophical cognition of the infinity of matter.[6] The position that together with the refutation of the identity of infinity and boundlessness the infinity of matter is also refuted, in no way agreed with the materialist dialectical view. A specified philosophical statement arrived at by the knowledge of a certain time proved untenable. But this did not apply to the general philosophical statement about infinity, in the sense of the inexhaustibility of matter. If space and time are forms of the existence of matter and if matter is inexhaustible in its relations, then this results, also in the potential inexhaustibility of real objective space-time structures which are continuously better understood by theory in their qualitative variety and in their internal unity of types and forms of matter.[7] The principle of the inexhaustibility of matter expresses above all the epistemological character of materialist dialectics not to recognize any ultimate truths about the structure of matter, that is, about the entirety of the forms of structure, motion and development proper to types of matter.

2.3.2 The Principle of Structuring of Matter

In the history of thought there have been many attempts to construct unified theories of the world. Thus, it is possible to conceive the world as the alienation of an absolute idea, as a non-ego created by the ego, as a fateful deployment of what is preformed, or as an entirety of elementary modules with a unique predetermination of all states. Theoretical development in natural science attempted to utilize the philosophical idea about the material unity of the world in unified theories. If, in this context, the qualitative variety of forms is reduced to one, and say, if the emergence of new qualities and the development of higher qualities is discarded, as is the case in mechanical determinism, or if the types of matter are reduced to identical qualitative, final indivisible particles, then theoretical difficulties arose and continue to arise. The unity of the world consists in its materiality, that is, objective reality has direct or indirect effects on our sense organs and these effects are experimentally analyzed and theoretically synthesized. Sometimes, Engels's remark on this problem is forgotten. Engels indicated the limits to our cognition of structure by stating: "At any rate, existence is an open question from the limit where our range of vision stops."[8] The insight about the unity of the world in materiality is a historical and a systematic research programme to prove this unity by the "long and protracted development of philosophy and natural science,"[9] and in order to develop general theories which uncover genetic and structural interrelations between formerly separate fiields. Structure is the entirety of the general and the specific, of the essential and the inessential, of necessary and accidental relations between the elements of a system in a certain period of time which is determined by the basic quality of the system. Structure is both coagulated development and structure in a process. Materialist dialectics opposes a static view of structure, but highlights the importance of investigations of structure. The principle of structuring of matter says: There exist relatively self-contained systems of objective reality which co-exist and interact, which develop lower or higher and which are determined by system laws in their essential patterns of behaviour.

2.3.3 The Principle of Dialectical Determinism

Dialectical determinism is the recognition of the conditionality and causation of the objects and processes in interrelation with other objects and processes.[10] Dialectical determinism is directed against the mechanical determinist view on the causality of events, according to which objective events are to consist of unequivocal dependencies of future states on present ones, which implies predetermination and the possibility of absolute prediction. In this manner, the variety of causes existing in the teaching of Aristotele of *causa materialis, causa formalis, causa finalis* and *causa efficiens* are limited to the aspect of the latter. Above all, mechanical determinism denies the objective existence of chance and of freedom. Dialectical materialism abolishes the understanding of a causal relation as the necessary arousal of a certain effect by a certain cause. Causal relations are direct and concrete mediations of the objective interrelation, and in this context, changes of state in the system (effects) arise through actions on a system (causes). The general causal principle as an integral part of dialectical determinism insists that all effects are caused, but not that a certain cause necessarily brings about a certain effect. This is possible only in borderline cases of interaction. Because we are always concerned with interactions, that is with a complex of causal relations, the necessity or chance of a causal relation is only determined by its place in the complex of relations. Causal analysis is therefore not simply the uncovering of a cause which necessarily produced an effect. A philosophical analysis of quantum mechanics showed the objective existence of chance as a constituent element of physical theory formation. Biology and psychology always raised objections to mechanical determinism. But the theoretical problem is more profound. The point is a determinism which connects objective chance and freedom with objective causality.

In discussion with criminologists about the theoretical foundations for proving objective causality, a general problem range of the dialectical understanding of cause became apparent.[11] A description of all causal relations which can be registered is not sought, but rather the initial cause for the final effect. In the concrete cases we investigated these were, for instance, the cause of death of a crane driver who did not comply with labour protection regulations or heavy injuries suffered by a motorcyclist riding without a driving licence but with a friend's knowledge on his motor bike. The caustic remark of a colleague was that each participant in public road traffic was a potential killer because his (even unpremeditated) action created the reasons for a later effect. Causes are identified with conditions and the causal course is viewed as an unavoidable causation of the effect. It share be explained that modifications of sequences of causes are possible by independent, conscious activity. Above all, in each case an analysis of the complex of conditions proved necessary because the conditions determine a field of possibilities. In this context the final effect is one of the effected possibilities from this field. Death and enjuries were not necessarily bound to occur. As an initial cause of a final effect only that complex of factors can be valid which puts into effect the possible final event with necessity or with great probability.

This also leads to an exact definition of the understanding of necessity. Necessity is often understood as a set of causes which make an event occur. If we proceed from the inexhaustibility of matter, then these conditions cannot be grasped fully. There exist random chances, that is a concomitant series of events which do not mutually justify each other.[12] If, therefore, necessity is defined as "determined on the basis of the set of conditions", then this is either an expression of mechanical determinism, if the exhaustive cognition of these conditions is principally asserted, or a programme for differentiating these conditions as necessary and accidental, essential and unessential. In this respect there emerge two types of necessity, that is, the simple direct necessity in which one certain cause unequivocally produces a certain effect, such as taking of poison in sufficient quantity to lead the death of a human being, and the general necessity, that is the reproducibility of essential aspects of an event in a certain complex of conditions. The first form of necessity is a borderline case of a causal relation between an initial cause and an unequivocally determined final effect, the second form leads to the cognition of the objective laws in theories because the law is universally necessary, that is a reproducible and essential interrelation which determines the character of the phenomenon.

Dialectical determinism distinguishes between causality and law. Causal analysis leads to the cognition of the conditionality of phenomena by differentiation between a complex of conditions and by insight into the initial causes and final effects. This causal analysis leads to the formulation of scientific laws as an expression of the determination of a phenomenon in its universally necessary and essential aspects.

2.3.4 The Principle of Development

The principle of development occurred in various forms in the history of thought. There exist the intuitive insights of the Greek philosophers. Circulation theories were put forth. Insights of the 18th and 19th centuries promoted the thought of development. Kant's idea about the evolution of the cosmos and Darwinism have substantially influenced the discussion about development. It is certainly interesting that Johann Georg Walch's philosophical lexicon which appeared in 1775 in its fourth edition does not contain the concept of development (evolution).[13] In this respect, F. Mauthner stated at the beginning of this century, "Whoever has sharpened his ears for scientific catchwords, will hear the struggle between purpose and chance from the all-explaining concept of Darwinism: from the word development... which then no more than fifty years later has developed and unfolded a vast literature on development."[14] Today, philoshopical discussion has been broadened about the development theory by new insights about the evolution of the cosmos, about bases in molecular biology for biotic evolution, about the theory of dissipative structures, about anthropogenesis, about the evolution of the Earth, and so on. The task is being solved which was posed by Lenin: "The continuation of Hegel's and Marx's work must consist in the dialectical treatment of the history of human thought, of science and technology."[15] In this context, the notion of development has developed. Materialist dialectics discusses circulation theories which deny the emergence of the new and it views critically the identification of quantitative or cumulative growth without considering the changes of quality. The Marxist materialistic dialectical principle of development comprises more than only qualitative changes, that is, the existence of other properties of the same fundamental quality and the emergence of new fundamental qualities. Development is the genetic interrelation between an initial quality and a higher final quality in a cycle of development, where the tendency towards higher development imposes itself through regressions, stagnations and through the formation of elements of a development phase. Thinking in terms of processes will then become thinking in terms of development, if in development cycles the higher qualities are determined by means of development criteria, if the objective dialectical opposites are analyzed in their deployment, solution and new definition as a source of development, and if the direction of development is recognized in dialectical negations of the negations as a unity between a seeming return to the old on the one hand, and a better qualitative fulfilment of the functions of the initial quality by the final quality, on the other.

The philosophical discussions about the relation between law and opposite show in an increasingly profound manner that by a deeper penetration into the structure of a law, laws prove to be tendencies for resolving objective dialectical opposites. But this must still be explained.[16]

Lenin spoke about the necessity of connecting the principle of the unity of the world with the principle of development. We are concerned with an explanation, how higher qualities are formed of lower ones (genetic interrelations) and which interactions emerge between them at certain stages (structural interrelations). This connection between the principle of determinism and the principle of development is effected in materialist dialectics above all by works on the concept of statistical laws and on the structure of development laws.[17]

2.4 THE TRUTH AND VALUE OF SCIENTIFIC INSIGHTS

In the synthesis aimed at social practice of scientific results by philosophy, the relation between truth and value and therefore orientation plays a decisive role for the moral behaviour of a scientist. Since the application of the atom bomb by the American imperialism, the discussion about the value of insights from natural sciences for Man has continually increased in scope. Bourgeois ideologists state crisis phenomena and develop pessimist or illusionary optimist visions for the future of scientific and technical development. Idealism penetrates into the thinking of many a scientist via ethical problems. It seems, as if the shortage of raw materials, the energy crisis, the population explosion and environmental pollution lead to a rift between Man and Nature. For bourgeois ideologists, a general crisis of Man is becoming manifest here. In this is expressed the consciousness of crisis inherent to imperialism as an essential feature. Ecological crises always existed, as is shown by the devastation of regions. Owing to an extensive utilization of science and technology, today the danger of ecological crises is greater. Decisive in this context is the deformation caused by the craving for profits, in the development of productive forces in imperialism, which results in many crisis phenomena which are no general crises of mankind, however. The cooperation of states with a different social system is possible and necessary, in order to utilize science and technology for the protection of natural conditions for human life. In this respect, the socialist line aimed at a conscious construction of an environment conductive to human beings is opposed by the imperialist line aimed at a profit-oriented exploitation of the natural resources. But imperialism is also interested in preserving the natural conditions for the production of profits, and environmental protection is increasingly becoming a profitable business.

The purposeful changes of genetic material, experiments with and on humans by transplantations of organs, by the influence of drugs on the structure of personality lead up to the question of an ethical moral foundation of biological genetic and medical research. There are added to this, discussions about the interruption of pregnancy and euthanasia. A struggle must be waged for orienting the research and utilization of scientific insights towards assistance for human beings. This must not be impeded by too narrow-minded moral demands. Too broadly viewed norms, however, can allow antihumane experiments. Such demands, as low risk, the social significance of research projects, suitable security regulations, education for responsibility and freedom for decision must be taken into account. The deeper penetration into the laws of processes affecting consciousness invigorates the demand for an ethics adequate to scientific development.

In order to meet the demand of obtaining humanists and materialists among the natural scientists as allies in the struggle against anticommunism, antihumanism and idealism, it is necessary to formulate our philosophical standpoints clearly.

Firstly, research in natural sciences is not generally free of values. Values are socially relevant facts, objects and insights which determine the material and cultural living standards of human beings in the concrete historical conditions of Man's coming to grips with his natural and social environment. Hence, values are relations between three elements, because the relations exist between fact, objects and insights on the one hand and Man, on the other, in the conditions of social practice which is substantially determined by the conditions of production. Therefore, it is not possible to agree with J. Monod, if he understands truth as the only value of natural science.[18] Truth is a two-digit relation between the fact and its image and does not take into account that meaning of true insights which is contained in the value. The task of our scientific investigations is to uncover the truth about the value by investigating the social development laws in the present-day conditions in their mechanism of action and in this manner, the relative aims (values) of further social development are determined. As humanists we orient the social forces at a humane and peaceful utilization of scientific insights, at establishing humane human relations. In this respect, peace, progress and freedom are basic values of socialism.

Secondly, the orientation of scientific research and the social utilization of its results are not primarily dependent on the ideas of value of scientists, but on the conditions of production which determine the ideas of value substantially. In imperialism, the dominating ideas of value which are written down in juridical norms and are spread as public opinion by the mass media are the notions of the ruling class which result from the monopolist ownership of the means of production and from the profit interests. It is not natural science and technology that have destroyed the moral foundations of civilization, as is stated by M. Born and many members of the Club of Rome. The revaluation of values demanded by them is necessary, because imperialist conditions of production make science a destructive force. Scientific potential is used to satisfy profit requirement, which leads to continual phenomena of crisis which cannot be abolished by other ideas of values. The real revolutionization of social conditions cannot be replaced by the imaginary revolution in ideas of value.

Thirdly, theoretically founded and practically feasible values for concrete social conditions are the theoretical basis for norms of behaviour which regulate activity and represent a concrete scale of values. In this context, it is necessary to take into account the interrelation between the theoretically founded imperative sentences or statements and the norms that are formed spontaneously. Both must be mutually agreed, if significant social norms are to regulate action. The interruption of pregnancy is legally allowed in this country. In this manner, the freedom of decision for women is guaranteed. The behaviour of physicians and of the environment towards a mother who refuses to have a child is determined by their ideas of value, which can lead to conflicts in the case of non-agreement. The investigation of the relation between socially recognized and sanctioned values on the one hand and traditional ideas of value on the other, a relation which can be expressed in various norms and leads until a contradiction between imperative sentence and spontaneously emerging norms, must therefore be carried out for the development of science in concrete terms for resolving the problems mentioned above. Thus, in my view, a legal regulation for the possibility of euthanasia is to be rejected on the grounds of humanistic considerations, which, however, does not eliminate the problem range of the wellfounded decision in terms of conscience for concrete cases.

Fourthly, there are long-term laws of scientific development, according to which in case of a social demand and of a suitable experimental and theoretical level, a personality is found who solves the problem involved. Heisenberg emphasizes the moral responsibility of inventors, but he thinks that by these laws he can affirm the replaceability of individuals in discoveries from which there results that the discover cannot assume full responsibility for technical implications. The problem range of the multivalent utilization of scientific insights must be absolutely taken into account, but it does not abolish the political and moral responsibility of the scientist. He must not search for truth alone, but must become aware of the humanistic meaning of his life, and this meaning consists in advocating peace, freedom and progress.

Marxist-Leninist philosophy and ethics must deal with the moral problems and problems of world-outlook in our time which comprise the relation between the truth and value of insights in natural science and therefore the answer to the question about the responsibility of the scientist in the present. For various conditions such solutions must be looked for which ensure social progress and show the humanist way. As realistic optimists who can cope with the difficulties, we oppose pessimism and illusionary optimism advocated by bourgeois ideologists regarding the social value of scientific and technical advance.

In this respect, the ideologically founded practical discussion must be held together with the humanistic scientists, in order to show political instead of military solutions of conflicts and to create social conditions, in which te antihumane utilization of scientific insights will be eliminated to a maximum.

By our theoretical work, values must be defined in concrete historical terms and the norms necessary for this must be deduced. This will only be necessary, if we make more profound the analysis of social laws, so as to recognize better the truth about the relations between objective laws and values, as well as to investigate more deeply the relationship between objective laws and social activity. The norms of our behaviour towards the development of science must comply with the possible objective development trends.

In this manner, scientific bases are created for responsible free action. Freedom is the mastering of the natural and social conditions for human existence in concrete social conditions by a type of action which is in line with competent decisions. Competence means knowledge of scientific truths and of social values. Thus the synthesis function of philosophy also has practical importance.

REFERENCES

1 H. Hörz, *Marxistische Philosophie und Naturwissenschaften*, Berlin 1976,
2 Marx-Engels, *Werke*, vol. 20, Berlin 1973, pp. 45–50.
3 H. Hörz, *Materiestruktur*, Berlin 1971, pp. 283–298.
4 H. von Helmholtz, Philosophische Vorträge und Aufsätze, editors: S. Wollgast; H. Hörz, Berlin 1971, p. 187 seq: cf. Introduction by the editors.
5 N. G. Tschernyschewski, *Ausgewählte philosophische Schriften*, Moskau 1953, pp. 668–669.
6 H.–J. Treder, *Relativität und Kosmos*, Berlin, Oxford, Braunschweig, 1968, p. 60 seq.; cf. also the works by W. A. Ambartsumyan, D. I. Naan and A. A. Selmanov.
7 Materialistische Dialektik in der physikalischen und biologischen Erkenntnis, editors: H. Hörz; U. Röseberg, Berlin 1981.
8 Marx-Engels, *Werke*, vol. 20, loc. cit., p. 41.
9 Ibid.
10 H. Hörz, *Der dialektische Determinismus in Natur und Gesellschaft*, Berlin 1974.
11 W. Griebe–H. Hörz–A. Lutzke, "Schöpferische Anwendung der marxistischen Philosophie auf die Kausalität im Strafverfahrenn, in: *Neue Justiz*, issues 23 and 24/1968.
12 H. Hörz, *Zufall–eine philosophische Untersuchung*, Berlin 1980.
13 I.–G. Walch, *Philosophisches Lexikon*, Leipzig 1775.
14 F. Mauthner, *Wörterbuch der Philosophie*, München, Leipzig 1914, Band I, p. 283.
15 V. I. Lenin, *Werke*, Bd. 38, Berlin 1964, p. 137.
16 H. Hörz – K. F. Wessel, *Philosophische Entwicklungstheorie*, Berlin 1983.
17 V. I. Lenin, *Werke*, Bd. 38, Berlin 1964, p. 242.
18 J. Monod, *Zufall und Notwendigkeit*, München 1971.

3. The Integrative Function of Philosophy

A. D. URSUL

Scientific philosophy plays an important part in the development and consolidation of the integrative processes going on in the realm of cognition, and also in the intensification of the interaction between the different areas of science and its subdivisions. As emphasized by I. T. Frolov, "in many cases, it is precisely philosophy that proves to be the focal point around which scientific forces can be effectively united and complex problems brought up for lively discussion... it is precisely philosophers who, in present-day conditions, should take up the initiative and become trail-blazers in organizing a more comprehensive union which, on the basis of the dialectical-materialist outlook and methodology, should bring together scientists representing natural, technical and social sciences".[1]

Philosophy's integration potential is most marked in the utilization of the principle governing the universal connection of things and that of the world's material unity. Other principles, laws, and categories of philosophy play a substantial role, too, in integrating the sciences of nature and society insofar as they can be used to identify the essential interconnections between the attributes and properties of matter and consciousness at a universal level. The very subject of philosophy and its dialectical-material method have an integrative general scientific content, just as the basic question of philosophy which investigates the relation of thought to being and the connection between them.

Integrative functions are also performed by science teaching on the forms of motion of matter and their interconnections, on which F. Engels based his systematization of branches of knowledge and which has been further developed by B. M. Kedrov,[2] in application to modern science and especially natural science.

What is of basic importance for the process of scientific knowledge integration is, of course, the interconnection of and the interaction between, the forms of motion of matter, especially in nature and society. It is precisely this interconnection that should, according to K. Marx and F. Engels, find reflection in the development of scientific knowledge: "We know only a single science, the science of history. One can look at history from two sides and divide it into the history of nature and the history of men. The two sides are, however, inseparable: the history of nature and the history of men are dependent on each other so long as men exist."[3]

Considerable integrative possibilities for cognition are inherent in the principle of the universal connection between things and that of the material unity of the world. "The whole of nature accessible to us", wrote F. Engels, "forms a system, an interconnected totality of bodies, and by bodies we understand here all material existences... The fact that these bodies are interconnected is already given in that they react one on another, and it is precisely this mutual reaction that constitutes motion".[4]

The reflexion of this interconnection in our consciousness leads to the synthesis of knowledge, i.e. the results and forms of cognition. However, under the impact of the objectively existing connections, the integrative tendencies tend to be reflected not only in the results of our cognitive activities but also in the latter's methods as determined by the subject of investigation, and also in the goals of the activities, and operations.

Viewing science as an activity (see further on), we cannot limit ourselves solely to the results that should adequately reproduce the object of cognition, and, hence, its links with other objects and the interactions within it. But if the given result of cognitive activities in the form of knowledge does, to a certain extent, reflect the objectively existing connections, then the other components of cognitive activities do reflect it in intermediate ways.

In materialist dialectics, the principle of the material unity of the world reflects the actual unity of matter, and this reflection consists in identifying the connections between material objects and their interactions, and also of their general, identical moments, laws, properties, etc. It is quite clear it is not philosophy alone but science as a whole that is engaged in disclosing the connections and common properties of objects, but as far as nature is concerned it is, above all, the concern of natural science with philosophy focussing its attention solely on the most general connections and relationships.

When F. Engels indicated that "the real unity of the world consists in its materiality, and this is proved not by a few juggled phrases, but by a long and wearisome development of philosophy and natural science",[5] he also meant that the problem of the material unity of the world was not only a philosophical one for it could be solved only by a united and integrated effort of philosophy and natural science. In modern conditions, when the union of philosophy and natural science has been joined by some particular sciences (and other scientific problems), too, and has thus become wider, the material unity of the world can be proved more successfully, because philosophy is helped in this by the social and technical sciences. Although the need to solve the problem of the material unity of the world in a complex fashion has not been the only reason for the union between philosophy and natural science, the broadened framework within which philosophy now maintains its allied relations with other sciences does raise the important question of a wider interpretation (not exclusively naturalistic) of the material unity of the world — and the participation of the social and technical sciences in providing proof of this unity is thereby made indispensable. Of the natural sciences, proof of the material unity of the world was until recently provided mainly by physics, but such proof is now furnished by chemistry, biology, astronomy, etc. as well.

The development of the scientific and technological revolution has resulted in the emergence of what may be called the integrative general scientific (and not only scientific) trends in research and has substantially consolidated the process of providing proof of the material unity of the world.

The integrative trends in research, such as cosmonautics, the ecology of man, the general systems theory and the systems approach, cybernetics, and others, have brought to light new facets of the material unity of the world, but their emergence has been made possible, in turn, by the presence of those aspects of the material unity of the world reflected in their subject-matter and results. Cybernetics has laid bare the most common features and interconnections of information and management in the biological and social forms of the motion of matter; systems research — the interconnection, integrity and hierarchy existing in nature and society; as for cosmonautics, it has further extended proof of the unity of the terrestrial and the cosmic, work on which was, in fact, initiated by astronomy; it has discovered new kinds of matter in the Universe and new inteconnections between the Earth and outer space, including the establishment of the interaction of the new anthropogenic factor in outer space which gradually results in the emergence of the largest socio-natural "Mankind — the Earth — outer space" system.

A demonstration of the material unity of the world has ceased to be the privilege of physics (and philosophy) and the sciences about nature in general. The social and technical sciences have also actively joined in this process. The dependence of nature on man, of which we are all aware today in the form of the outcome of man's economic activity on our planet, has demonstrated that material unity is not purely natural in its character; however, attention has been concentrated on it until recently. The material unity of the world in those areas where man interacts with nature cannot be disclosed by the sciences of nature alone, because society interacting with nature also represents matter, at the highest stage in its modern development. The process of providing proof of the material unity of the world is not joined in only by integrative scientific directions but, *via* them, by the social and technical sciences as well. The technical sciences reflecting the laws governing the motion of the material means of human activities, which represent the connecting link in man's interaction with nature, are from the broad world outlook standpoint involved in proving the materiality of the means of human activities whereby nature is recognized and transformed and protection against pernicious environmental effects is ensured. The technical means of human activities bring what is natural and what is social together and thus represent the connecting link between these two qualitatively different and interacting states of the motion of matter.

Thus, we find that all the main subdivisions of modern science have joined forces in providing proof of the world's material unity and that argumentation is not confined solely to the sphere of a certain science or realm, it goes on in the border lines between scientific disciplines and subdivisions in the process of their interaction. Now we can say that providing proof of the material unity of the world is not only the concern of philosophy and natural science, but also of science as a whole, it has now become a task of general scientific importance, calling for a more involved interaction between social, natural and technical sciences. In this sense, the intensification of the integrative processes in science and the rapid growth of general scientific forms, tools and trends in the realm of cognition signifies a further reflection of the material unity of the world, and the fact that obtaining further proof of its materiality is now the concern of the entire system of disciplines, i.e. science as a whole rather than just some parts of it. This is a new stage in the synthesis of the knowledge of nature and society, of which K. Marx dreamt when he spoke about a single science.[6]

Among the factors and sources of integration going on within science with the aid of philosophy as an activity, one can identify onthological, epistemological, logical, etc. Of special interest to us are the logical-epistemological factors of integration, which are connected with cognition of the objective truth, with the movement of knowledge towards the absolute truth. In the course of this movement, what takes place is not only a mere accumulation of truths already discovered by mankind as if they were bricks in an edifice, but their change, development, rejection of absolute notions and what is the main thing — the synthesis of relative truths into an integral system of the knowledge already available, a system which is fixed not only in the archives of various disciplines but, above all, in the scientific picture of the world, the scientific world outlook and the scientific philosophy. The progressive change in the systems-*cum*-truth states of scientific knowledge mainly represents the process of knowledge integration because the truth, as was emphasized by

V. I. Lenin, is composed of "the totality of all sides of the phenomenon, of reality and their (reciprocal) relations".[7]

Since Marxist-Leninist philosophy does, to a considerable extent, utilize the generalizations and the synthesis of knowledge of particular sciences and thus, in performing its methodological functions further in the course of cognitive activities, it implements the synthesis of knowledge — both the synthesis of knowledge in individual scientific disciplines (internal in relation to the given individual scientific discipline) and the "external systhesis", i.e. the synthesis between various diciplines and their main subdivisions. The point at which the main "wave of integration" occurs is, evidently, at the interface between philosophy and particular sciences in that type of synthesis of knowledge, which according to B. M. Kedrov can be called a "synthesis *via* philosophy".[8]

The interaction of philosophy and natural science, and the union of philosophers and natural scientists have, for a long time, been playing the leading role in synthesizing scientific knowledge. In recent years, what has been taking place is not only a further consolidation of allied relations between the philosophical and natural sciences but also a widening of this union, and its spread to other sciences. As for the union between philosphy, on the one hand, and the social sciences and the humanities, on the other, we should like to note that such a union in the form of integral teaching, and namely the Marxist theory comprising materialist philosophy, political economy, and scientific communism, has been in existence since the very rise of Marxism and has been exerting its influence upon the development of the social sciences ever since.

Historically, it has so happened that the intensive development of philosophical research in the field of natural sciences signified a substantial contribution to the development of the union of philosophers and natural scientists and now the forms and experiences of their fruitful relationship have been spreading to all other sciences. But, genetically, the development of allied relations with social knowledge (political economy and scientific communism), should be placed first. At the present time, the comprehensive union between scientific philosophy and all the special sciences, is being developed in various forms, including joint research by philosophers and representatives of special sciences, and also the organizational forms of development of their interaction, such as holding philosophical (methodological) seminars in scientific establishments, participation in the work of the respective scientific councils, conferences, symposia, regional, all-Union and international congresses, etc.

It is obvious that in such a case we view philosophy not as a certain system of knowledge and even a form of social consciousness but from a different angle — as phenomenon of culture and a specific type of intellectual activity of people. The activity approach is applicable to philosophy, too, and philosophy can also be represented as a definite kind of human activity, interconnected with other types of activity both within science and outside it, above all, in, the fields of social management, political, ideological, propaganda, pedagogical organizational activities, etc. Such a view of philosophy as a kind of scientific and social activity, which corresponds to the actual position of the philosophy of Marxism in a socialist society and its growing role in modern world,[9] calls for its detailed elaboration. One of the reasons for the need to develop the activity approach to philosophy is its increasingly important role in the integration of the sciences, in strengthening the interaction of the social, natural and technical sciences.

Among the various conceptual notions of science the most general, is, in our view, the interpretation according to which science is viewed as a special type of activity in the realm of producing new scientific knowledge and the latter's utilization in various spheres of its practical application. As a specific form of human activity science includes the following interacting components: subject, object, means, requirements, objectives, operations, conditions, results (depending on the aspect of investigation by the science, a number of the main components are subject to change). Apart from the systems-component aspect, science which is itself a specific type of social activity also includes the following aspects: scientific research, scientific organizational, scientific informational, scientific-pedagogical and scientific supporting. These components and types of activity making up an integral system exist in every branch of science.

The activity approach to science does not contradict the traditional reflective (epistemological) approach, but it does, to a certain extent, embrace it, complement and render it concrete, focussing attention on the subject and the subject-object relations, and this links the epistemological and the sociological approaches to science into a single whole. The systems-activity approach offers the advantage of a more fruitful methodological analysis of the differentiation and integration of science, the analysis of the trend to achieve the unity of science. If in the case of the reflective approach attention is focussed on the subject, method and results of individual branches of knowledge, and the problem of differentiation and classification of science is considered mainly from this point of view, the systems-activity approach offers an opportunity of increasing the number of components, discovering in more detail the specific features of science as an activity. The activity approach to the differentiation of science allows the process of specialization to be viewed as a definite "mechanism" responsible for the growth of the effectiveness of cognitive activities if effectiveness is understood not only as economic effectiveness (ratio of result to cost), but in a wider sense as the outcome, the effect of scientific activities (including various kinds of ratio of result to the requirements, objectives, quality, etc.).

However, general scientific integrative processes which are connected dialectically with differentiation also pursue the objective of raising the effectiveness and enhancing the outcome of scientific activity, and it was precisely this point to which F. Engels turned his attention in his "Dialectics of Nature" when he emphasized that it was precisely at the point of contract between the sciences that "the greatest results are to be expected".[10] In the final analysis, no matter whether it is a case of differentiation or integration, they are pursued to raise the socio-economic effectiveness of science as a specific system of activity within the total system of human activities, to stimulate a further growth in the effectiveness of social production and improve the quality of work at all stages of human activity.

The integrative processes in science can be viewed, above all, as integration of the components of activity, i.e. of the subject, object, means, purposes, results, etc. Besides, the continuing integration also causes consolidation of the interconnection between the major and minor subdivisions of science, i.e. between natural sciences, social, technical, medical, agricultural and other sciences, between scientific research, scientific organizational, scientific informational and other types of scientific activities, between fundamental and applied, theoretical and experimental research, etc.

The activity approach to the integration of science allows the role of ontological, socio-cultural, logical-epistemological, psychological, ethical, aesthetic and other factors in the development of synthesizing tendencies in cognition to be thought out in a more profound manner, and the peculiarities and ways of moulding a single science, presupposing, rather than excluding, its developed differentiation and specialization, and a further discipline orientation to be disclosed to a greater extent.

The activity approach makes it possible to consider allied relations between philosophy and particular sciences as well as other areas of scientific quest not simply as an interconnection between two kinds of knowledge, but as an interaction between processes and types of activity, including as an indispensable component its results, i.e. the respective systems of knowledge. The methodological fruitfulness of such an approach is already obvious but further collective research in this direction, elaboration and recognition of the synthesis of traditional and activity approaches in science is still required.

The union between philosophy, on the one hand, and the three main branches of modern science, on the other, tends to consolidate the interconnection between them, serves as an important source of the development, diffusion and intensification of general scientific integrative processes.

Representing the nucleus of such general scientific integrative forms of cognition as methodology, outlook, the scientific picture of the world, philosophy exerts a tremendous influence upon the coming together of the sciences about nature, society, and the development of technology both directly and through these forms of cognition. The powerful integrative impulse coming from philosophy as the world outlook, as methodology and a more general picture of the world, creates quite a favourable intellectual climate, and mental conditions for the growth of integrative tendencies in other branches of knowledge and between them, promoting the emergence of new general scientific processes of which we shall speak below.

Until recently, the new phenomena of a general scientific nature have not been accorded the attention they deserve. This can be explained not by the non-existence of general scientific forms and tools of cognition at the previous stages of scientific development, for they always existed in one form or an other, but by the fact that they simply did not play such an important role in the development of science as they do in the epoch of the scientific and technological revolution.

The general scientific components of knowledge, for example, the methods of abstraction, idealization, analysis and systhesis, induction and deduction, logic and historical methods, etc. had been known previously and did exist prior to the epoch of the scientific and technological revolution. However, they did not play the role they do today and their development did not reach a sufficiently high level to attract the attention of philosophers to make an independent study of them.

It was precisely the intensification of the integrative processes in science that is responsible for the appearance of qualitatively new general scientific epistemological phenomena, and it is precisely these new forms and tools of cognition that have emerged in the course of scientific integration in the conditions of the scientific and technological revolution that will be the object of our concern here. Among the new phenomena of general scientific importance in the system of knowledge one can identify such concepts and categories as information, system, model, symmetry, uncertainty, and others; methods and approaches — modelling, cybernetic, informational, etc., the complex problems relating to the main groups of sciences (the conquest of outer space, developing the World Ocean, the problem of man, etc.). Further, we shall dwell only on the role of philosophy in shaping those new general scientific integrative forms and tools of scientific cognition, isolating what, in our view, constitutes the most principled problems, without touching the entire range of problems which were discussed in other works.[11]

The study of general scientific integrative phenomena makes it possible approach the logical-epistemological structure of the science as a whole from a new point of view, from positions of the relationship between philosophy and the particular sciences. The dichotomic division of scientific knowledge would seem to change into a three-pronged affair in accordance with the formula "general-specific-single" (philosophical — general scientific — particular scientific). But in reality, what we have in mind is a large number of components and forms of scientific cognition which are located between philosophy and the particular sciences and also within them, and, apart from traditional spheres of knowledge. Transitions from philosophical concepts (categories) to particular scientific forms and tools of cognition prove to be smoother and less discrete than was heretofore presumed.

The development of science in the second half of the twentieth century is a complicated, manifold, and contradictory process, accompanied by the appearance of new integrative phenomena and tendencies, among which the most essential and uncommon ones are general scientific theoretical forms and tools of cognition. Their appearance is evidence of the historial nature of theoretical cognition, fresh confirmation of the proposition of the classics of Marxism to the effect that "in every epoch... theoretical thought is a historical product, which at different times assumes very different forms and, therewith, very different contents".[12]

The development of general scientific theoretical forms and tools of cognition on the way to the unity of scientific knowledge is evidence of not only the rapprochment of the various particular scientific disciplines, but also a symptom of their intensifying interaction with philosophy. In a definite sense, one can say that general scientific knowledge is a new facet in, and even a stage of, the realization of the union of the philosophy and the special sciences, the union which has given birth to a new offspring bearing in its character the features of both parents (not all features, of course, but only those that arose in the course of synthesis). Without rejecting or superseding the traditional forms of interrelationship between philosophy and the particular scientific knowledge, the union acts as an additional form of interconnection between them, consolidating and promoting their mutually fruitful co-operation. In this phenomena, we see further progress according to the Leninist proposition on the relationship between philosophy and natural science because the formation of general scientific knowledge has become possible only in the conditions in which the three main groups of sciences interact with philosophy.

Sometimes, in philosophical literature, one comes across the thought, perhaps in veiled form, that the methodological role of philosophy in relation to special sciences is only belittled by the special setting apart of the general scientific epistemological phenomena. But if this thought does indeed occur in some works, this should not serve as the basis for "prohibiting" the separation and investigation of these phenomena, provided the leading role of philosophy in their emergence and development is duly taken into account. Our point of view is that the appearance and separation of general scientific knowledge tends to intesify and elevate the role of philosophy's integrative function rather than belittle it.

The status of the general scientific forms and tools of cognition has as yet been investigated insufficiently and, therefore, causes much argumentation, as is the case with any developing phenomena coming within the sphere of scientific quest, and the mental communication between scientists. The debatable questions apply, in our view, not to the principled separation of this kind of knowledge, but only to some of its types. General scientific non-philosophical knowledge has always existed, and, as was noted above, such traditional tools of cognition as analysis and synthesis, induction and deduction, idealization, generalization and many others, the principles and laws of formal logic have always had a general scientific significance. They have been made use of both by philosophers and specialists in other areas of knowledge ever since science came into being as a phenomenon of human culture.

Philosophy has exercised its influence on some sciences largely with the aid of these forms and methods of scientific investigation, while as a methodology, philosophy has always represented teaching on those tools and methods and about its own philosophical method of cognition.

The traditional and now new general scientific tools of cognition (concepts, such as information, model, uncertainty, symmetry, extremity, effectiveness, etc.; methods, such as systems — structural, informational, modelling, etc.), are in terms of their effectiveness, quite adequate tools for integrating scientific knowledge, and represent one of the important forms of the interaction between the natural, technical, and social sciences. The very fact that the general non-philosophical and philosophical tools of scientific quest are utilized by the three main subdivisions of modern science points to their integrative possibilities, to their active participation in the synthesis of knowledge.

The appearance of new general scientific forms and tools of cognition, moreover, substantially strengthens the interconnection between philosophy and the particular sciences because they signify the appearance and widening of the new "channel" of interaction with the traditional ways of exercising a mutual influence on those which remain as they were without being superseded.

The new general scientific forms and tools of cognition mainly arise in the course of scientific integration and, in their turn, promote the intensification of this general scientific process. At the same time, they are not unresponsive to the processes of science's differentiation. Their general scientific nature manifests itself only in cases when they spread to the traditional and also new regions and subdivisions of the science, thereby promoting the development of differentiating knowledge and permeating the most diverse branches of science.

The general scientific forms and tools of cognition, both the traditional and the new, mainly belong to the theoretical level of cognition (probably excluding such tools as computers, measurement, partial modelling), which manifest their universal status in the realm of empirical research as well. Such appurtenance may not mean any rigid connection with a theory that is general scientific in nature, though the existence of specific general scientific theories cannot be doubted at all. An example of such a specific theory among the traditional ones, which are general scientific in nature, can be provided by formal logic among the mathematical disciplines, by the theory of sets, among the new theories, by the mathematical theory of categories, the general theory of systems, a number of cybernetic theories, informatics, forecasting, etc.

But by no means all the scientific concepts are rigidly connected with a definite theory, for there are, in particular, empirical notions (having no general scientific nature), notions which do not belong to a scientific theory but to theoretical systems — the scientific outlook, the scientific picture of the world, the various hypotheses, concepts, ideas, etc. Therefore, the general scientific nature of knowledge taken as a characteristic feature of it (concept, method, principle, etc.), does not always mean that it is rigidly connected with some concrete theory although, as noted above, this may well take place. Since general scientific knowledge is in a state of continuous formation and development, it may leave one system of knowledge for another, one theory for another, figuring in fact, as an inter-theory exchange component. In modern cognition the general scientific forms and tendencies represent the most dynamic forms of nexus between the various branches of science, above all, social, natural and technical branches, which effectively promote the interconnection between them, while the appearance in recent times of a considerable number of new forms of general scientific knowledge provides evidence of the growing consolidation of the three main subdivisions of science.

A characteristic feature of many integrative processes going on in the field of cognition is their immediate or essential nexus with various tendencies assuming the nature of general scientific ones, such as dialectization, mathematization, cosmization, sociologization, economization, etc. which reflect individual aspects of the influence that is exercised by a certain discipline, problem, subdivision, fragment, and component of a science upon those of another science. Such an influence or mutual influence (to which, in the final analysis, the very logic of the development of science as a system of interconnected parts, which themselves at a new stage in establishing their unity should lead) expresses the essence of the integrative processes in science, above all, the essence of the interaction between the natural, social and technical sciences, the enhancement of their mutual influence which is leading to the greater efficiency of science as a whole.

Some scientists consider that once general scientific knowledge is used in many or all sciences it is identifiable with philosophical knowledge. But we cannot agree with this. In our view, philosophical knowledge differs from the foregoing general scientific forms and tools of cognition in respect of a number of properties. Further, we shall dwell on the most important of them.

First, what constitutes the main thing for philosophical knowledge is the fundamental question of philosophy — the relation between matter and consciousness, the subject and the object. It is precisely this relation that determines the specific features of the philosophical problematic range, which was amply demonstrated by V. I. Lenin in his work "Materialism and Empirio-Criticism". Lenin considered that the foregoing relation separates the direction of philosophical epistemological research from that of particular sciences, for example, physics.[13]

Beyond the fundamental epistemological question where the antithesis between matter and consciousness is relative, because it has its limits, as was pointed out by V. I. Lenin,[14] there lies not only what can be viewed as the particular scientific but also general scientific knowledge. This separates what is common to material and ideal objects, subject and object.

Second, philosophical knowledge is just as related to science as a whole and the latter's various branches represent an actual universal-cum-general scientific knowledge while the general scientific phenomena discussed here turn out to be general scientific only potentially; their general scientific status is still in the process of establishing itself.

It stands to reason that the general scientific quality of philosophy is also subject to development and enrichment, thereby acquiring a new content, but this is a somewhat different process, because what we have here is the development of the actually available general scientific status but already in depth. Philosophy manifests its general scientific status not only in science, but in the whole of society's mental culture, in the thinking of people, while the functioning of general scientific knowledge is as yet limited by the framework of science.

Third, the logical-epistemological limitedness of general scientific knowledge, compared to philosophical knowledge, also consists in the fact that the latter is used at all stages and levels for solving all scientific problems. The general scientific forms and tools of cognition, while showing the potential possibility of diffusion to all branches of science are not applicable at all stages and levels of cognition (in particular, utilization of a number of them at the level of philosophical science is not obligatory) and not at all stages of problem-solving, but only for a definite aspect of research relating to content, and the specific nature of general scientific phenomenon.

Fourth, general scientific knowledge, to certain extent, renders concrete some (far from all) philosophical propositions, and develops them at its own level, thus offering an opportunity to elucidate philosophical ideas somewhat. The development of a certain general scientific concept or method (approach) is influenced by philosophical methodology and represents the further concretization (this time general scientific) of a range of philosophical propositions.

It is in this sense that L. F. Ilyichev, considers that general scientific methods can be looked upon as "concrete manifestations of a dialectical materialist method in one sphere of scientific cognition or another... General philosophical methodology permeates the methods of the special sciences and is related to them as is the general to the particular".[15]

The relation of the universal to the particular, in this case, to the general scientific, is what constitutes the principled relation from which one should proceed in analyzing the relation of philosophy to general scientific knowledge.

When we say that general scientific knowledge is a definite concretization of certain ideas of philosophy, we do not mean that it is only a simple reproduction of general philosophical provisions. This knowledge is also shown to be under the influence of its own subject of research although under the determining influence of philosophical knowledge. In effect, what emerges is the general scientific explication of a number of philosophical propositions and, like any explication, it "captures" only part of the content of the *explanandum* (philosophical knowledge).

Fifth, while existing as a specific form of scientific knowledge, philosophy stands in certain opposition, as far as its content is concerned, to mathematical knowledge. Philosophical knowledge and, in principle, the major part of it is, to a considerable extent, not amenable either to formalization or mathematization; it exists, as it were, in a "dimension" different from that of mathematical knowledge, with the former determining all the other kinds of knowledge methodologically. To explain our thought, let us resort to an analogy with. Descartes' axes of co-ordinates where the axes are perpendicular to each other: this relation of "perpendicularity" is intrinsic to both philosophical and the mathematical knowledge which, in this sense, are neither "superimposed" on one another nor "replace" one another, but complement one another within the entire "expanse" of scientific knowledge.

At the same time, since it belongs to a different level (dimension) of abstraction than philosophical knowledge, general scientific knowledge does not stand in the same opposition to the mathematical tools of investigation. Moreover, it has been established that the majority or even all the general scientific epistemological phenomena, known today, have their own mathematical "paraphernalia" (the systems approach — different variants of the general theory of systems; information concepts statistical and other theories of information, etc.). The mathematical and formal tools of cognition make it possible to explicate many propositions in the content of scientific knowledge and, due to this, they represent an effective means of research for many branches of knowledge. The specific features of general scientific knowledge as considered above in relation to philosophy provide evidence of the former's relative independence in science, which is determined by philosophical knowledge within the framework of the whole of science.

General scientific knowledge is shaped not only under the dominating impact of philosophy but at the same time it represents that additional and rather powerful channel (which is far from being the only one) through which the process of dialectization of modern science is continuing to broaden. It would not be correct to imagine that the methodological impact of philosophy upon particular sciences is effected solely via general scientific knowledge. Philosophy may exercise its influence upon particular sciences not necessarily via general scientific forms and tools of cognition, but also directly and immediately. The traditional mode of influence as exercised by philosophy upon particular sciences, which was the only one before the emergence of general scientific phenomena, still remains the main one, but apart from that, another channel of philosophy's influence has also emerged which has substantially accelerated and intensified the flow of scientific knowledge in either direction — from philosophy to the particular sciences and back.

In specially setting the general scientific forms and tools of cognition apart from philosophical knowledge and that of the particular sciences, we wish to stress that the very moulding of these new phenomena (an objective process) and their realization (a subjective process) did and does take place under the guiding and shaping influence of scientific philosophy and represents one of the directions (and results) of its integrative function.

The latter is constantly growing in importance, especially as the "tree" of science is becoming more branched. The integrative function also manifests itself via the methodological and world outlook function because the realization thereof, while fully conforming to the requirements of scientific and practical development, unavoidably leads to the synthesis of knowledge, and the integration of sciences.

Apart from the relatively independent autonomous disciplines, which can be particular or special sciences, other fields of knowledge also arrive where philosophy immediately enters and also provides the source of their emergence. Such areas of knowledge are not autonomous in their native state while representing in effect integrative interdisciplinary phenomena in which philosophical and purely scientific knowledge are blended into a single whole, with the central role being played by philosophy. The task of enumerating all such areas of knowledge is not set here but it would be quite appropriate to note that, among them, such theoretical systems of knowledge as methodology, world-outlook,[16] and the scientific picture of the world are to be found. In principle, they have an integrative general scientific nature, and are shaped by all branches of science, but their nucleus is formed by scientific philosophy.

The establishment of the complex, integrative nature of methodology[17] is an important conclusion which dialectically solves the question of the status of methodology generally and the methodology of science, in particular. Methodology arose within philosophical knowledge but gradually as a result of the concerted action of philosophy and particular sciences has transcended its boundaries without, however, losing its immanent connection with it. Methodology has several levels but, in some way or another, they are not all simply connected with philosophical propositions, but are determined by them.[18]

This determining level of methodology relating directly to philosophy represents philosophy as a whole in that aspect thereof in which it is called upon to implement its methodological function in relation to all non-philosophical areas of knowledge. Philosophy's other level which is formed mainly in the sphere of particular sciences is represented by the most general basic propositions and principles of those sciences which predetermine the directions of scientific quest in those branches of knowledge.

One can also identify the level of methodology which is, to a certain extent, an intermediate one between philosophical and particular scientific levels, i.e. the general scientific level. Its appearance is the result, on the one hand, of the intensified interaction between the social, natural and technical sciences, and, on the other hand, the continuation of further integrative tendencies in methodology, of a definite synthesis of its philosophical and particular scientific levels. The question of the identification of the general scientific levels of methodology is a debatable one. It has begun to be studied comparatively recently, but, in our view, there is every ground for considering its separation completely justifiable. It is important to emphasize that this new intermediate level of methodology should be treated in the spirit of a concession to the pluralism of methodologies because the general scientific level of methodology, like the particular scientific one as well, is determined by the philosophical level of methodology.

The identification of the general scientific level of methodology turns out to be necessary for the further study of the methodology of particular sciences, like their regions or main groups of sciences, among which we place the social, natural and technical sciences. But it is the role of the general scientific level of methodology which is especially important in studying the process of the intensified interaction between the social, natural and technical sciences, which represents one of the main tasks of modern scientific development, raising the latter's effectiveness and influencing human activities and culture in their entirety.

REFERENCES

1 I. T. Frolov, "Nauchno-tekhnicheskaya revolutsiya i novaya rol filosofii" (The Scientific and Technological Revolution and the New Role of Philosophy), in: *Voprossi filosofii* magazine 1973, No 9, p.59.

2 B. M. Kedrov, *Klassifikatsiya nauk* (Classification of Sciences), book 1, Moscow book 2, Moscow, 1965.

3 K. Marx—F. Engels, *The German Ideology*, Progress Publishers, Moscow, 1964, p. 28.

4 F. Engels, *Dialectics of Nature*, Third Revised Edition, Progress Publishers, Moscow 1964, p. 71.

[5] F. Engels, Anti-Dühring, Second Edition, Foreign Languages Publishing House, Moscow, 1959, p.

[6] K. Marx—F. Engels, *Collected Works*, vol. 13, p. 497 (in Russian).

[7] V. I. Lenin, *Collected Works*, vol. 38, Conspectus of Hegel's Book "The Science of Logic", Progress Publishers, Moscow 1964, p. 196.

[8] B. M. Kedrov, "O sinteze nauk" (On the synthesis of Sciences), in: *Voprossi filosofii* magazine, 1973, No 3; see also: "Rol materialisticheskoi dialektiki v sinteze nauk" (The Role of Materialist Dialectics in the Synthesis of Sciences) by the same author in the book: *Filosofiya marksizma i sovremennaya nauchno-tekhnicheskaya revolutsiya* (The Philosophy of Marxism and the Modern Scientific and Technological Revolution) Moscow, 1977.

[9] L. K. Naumenko, "Dialektiko-materialisticheskaya filosofiya v sovremennom mire" (Dialectical-Materialist Philosophy in the Modern World), *Kommunist* magazine, 1979, No 14.

[10] F. Engels, *Dialectics of Nature*, Third Revised Edition, Progress Publishers, Moscow, 1964, p. 297.

[11] V. S. Gott—A. D. Ursul, *Obshchenaychniye ponytia i ikh rol v poznanii* (The General Scientific Concepts and their Cognitive Role), Moscow, 1975; V. S. Gott, A. D. Ursul, E. P. Semenuk, *O edinstve nauchnogo znaniya* (On the Unity of Scientific Knowledge), Moscow 1977); also: *Sintez znaniya i problema upravleniya* (The Synthesis of Knowledge and the Problem of Management), Moscow, 1978; E. P. Semenuk, *Obshchenauchniye kategorii i podkhodi k poznaniyu* (The General Scientific Categories and the Approaches to Cognition), Lvov, 1978; also: *Dialektika i obshchenauchnyie metodi issledovaniya* (Dialectics and the General Scientific Methods of Investigation), Havana, 1981.

[12] F. Engels, *Dialectics of Nature*, Third Revised Edition, Progress Publishers, Moscow, 1964, p. 44.

[13] V. I. Lenin, *Collected Works*, vol. 18, p. 274 (in Russian).

[14] V. I. Lenin, *Ibid.*, vol. 29, p. 90 (in Russian).

[15] L. F. Illyichev, "O metodologicheskoi funktsii istoricheskogo materializma" (On the Methodological Function of Historical Materialism), in: *Voprossi filosofii* magazine, 1977, No 6, p. 27.

[16] The integrative-complex nature of world-outlook and the central role of philosophy in world-outlook was disclosed in detail in the report by P. N. Fedoseyev at the 16th World Philosophical Congress (see P. N. Fedoseyev, *Filosofiya v sisteme mirovozzreniya*—Philosophy in the System of World-Outlook, Moscow, 1978) and also *Mirovozzreniye, filosofia, nauka* (World-Outlook, Philosophy, Science) by the same author, Moscow, 1979.

[17] The philosophical questions of natural science may provide an example of the interdisciplinary integrative nature of methodological research. It is in the course of this research that the mutual links between philosophy nad natural science are implemented. The philosophical aspects of natural science are elaborated upon by philosophers and natural scientists jointly; they cannot be referred to as exclusively philosophical or as belonging to the natural sciences, for they have a complex nature and promote the realization of the unity between philosophy and natural science.

[18] This point of view is propounded in the interesting article by L. F. Ilyichev "O sootnoshenii filosofskikh i metodologicheskikh problem" (On the Relationship between Philosophical and Methodological Problems), in: *Voprossi filosofii* magazine, 1976, No 4.

4. The Main Forms of Interdisciplinary Development of Modern Science

S. N. SMIRNOV

The phenomenon of interdisciplinarity in modern science is existing and developing in a vast diversity of different forms, ways and means of establishing and advancing of various interdisciplinary relations. On this strength we face the acute problem of general systematization and classification of the main forms of the phenomenon in question.

One could bring out a whole series of grounds for tackling the problem. But in doing so we must take into account that systematization and classification of the whole variety of interdisciplinary forms should be based on such a classificational foundation that could provide us with the possibility to reveal the natural, regular, governed by certain laws interconnection of different types and ways of interaction of the modern scientific disciplines. In such case we could be also in a position to single out the basic mechanisms of the intercommunication, intercommunion, intercross, interdependence, interfusion, intergrowth, interlacing, intermarriage, intermutation, interosculation, interpenetration, intervention and other processes of interdisciplinary relations between different scientific formations existing in the "body" of contemporary science.[1]

It seems essentially obvious that such a classificational foundation could be most successfully found in the general unity of various scientific disciplines,[2] in a single nature of scientific knowledge and cognition, in the common ways of existence of different sciences and in the same means of production, justification and social institutionalization of scientific knowledge.[3] Of course, it is necessary to keep in mind that interaction of the scientific disciplines is impossible not only without their unity, but also without their essential difference, irreducibility to each other. The interdisciplinarity takes up its stand upon the unity of qualitatively different areas of scientific cognition. That is why the classification of the forms of interdisciplinarity comes down to the systematization of the types, manifestations of the unity of sciences.

As known,[4] science is a specific social system of socio-psychological and experimental-theoretical production of system-organized knowledge on regularities existing between phenomena of the reality. Such a concept of science gives us the opportunity to single out the following fundamental types of the unity of scientific disciplines and of the corresponding basic levels of interdisciplinary forms of scientific progress.

In the first place, the unity of sciences can be an ontological objective unity of their subject-matters. In other words, it can take the shape of material connection of the fragments of the reality which have been investigated by the given scientific disciplines.

In the second place, the unity of sciences can manifest itself in common epistemological structures of different scientific "bodies" — conceptual apparatus, theoretical constructions, methods and principles of creation, generalization, systematization, theoretical and experimental substantiation of the scientific knowledge.

In the third place, the unity of sciences can incarnate itself in the same structure and functioning of the social institutes of different scientific formations, in essentially similar socio-psychological forms of knowledge creation and organization as well as management of this process as a specific social one, and, at last, in the common social and ideological forms of axiological assessment, acceptance and distribution of the knowledge produced by this or that scientific community.

This three types of unity of scientific disciplines establish the corresponding most fundamental levels of the forms of interdisciplinarity — ontological, epistemological and sociological (socio-institutional) ones.

Ontological, epistemological and sociological forms of interdisciplinarity are respectively independent from each other. But in the long run they inevitably interwork with each other so that their respectively autonomous existence, functioning and development result into closely interrelated moments of any sufficiently developed interdisciplinary scientific enterprise. Of course, interdisciplinary interaction of some sciences is able to begin at any of the three above-mentioned levels of interdisciplinarity. This notwithstanding cannot take the complete, mature form without acquiring some apt, fairly advanced forms relating to both other levels. For example, if a particular interdisciplinary formation arises from revealing some objective unity of one or another scientific discipline, then the complete, mature form of this interdisciplinary formation may be acquired by it. Only then the unity in question will begin to be discovered by conceptual, theoretical and methodological means which are adequate to the nature of this unity and also only then the production of knowledge on the unity will be properly organized at the sociological level, will be embodied in the social institutes which are adequate to the subject-matter and method of given interdisciplinary formation.

The fact that interdisciplinary interaction of sciences may begin at any of the three levels in no way signifies that they are all equal and have the same importance for the development of interdisciplinary researches. The most fundamental, determining form of interdisciplinarity is ontological. Neither epistemological nor sociological forms can exist without a certain objective unity of the subjects-matter of different scientific disciplines.

Singling out these three main levels in the forms of interdisciplinarity. first, poses a philosophico-methodological problem like the need to show the regularities of the interconnection between them, and second, signifies that classification of the concrete forms of interdisciplinarity can develop further as systematization of its various ontological, epistemological and sociological forms, bearing in mind that within this or that context of relations between the disciplines crucial importance may attach to this or that concrete aspect of interdisciplinarity which is either ontological, epistemological or sociological.

4.1 ONTOLOGICAL FORMS

4.1.1 Regional Concrete-Science Interdisciplinarity

In accordance with the division of the whole of scientific knowledge into fundamentally distinct (by their objects, above all) regions or areas of science like physics, chemistry, biology, the sciences of the Earth, astronomy, the social sciences, the sciences of man, the technical sciences, mathematics, philosophy there is a necessity to draw a distinction and make a special study of regional interdisciplinary processes above all within these concrete sciences. The point is that the unity of everything that exists is realised, one might say, in the form of "condensations" of greater unity linked with each other by a unity that is weaker than it is within these "condensations" themselves. The unity of nature and society is not as great as the unity of either nature or of society. The unity of physical and biological nature is not as great as the unity of either physical or biological nature. The unity of physical and chemical phenomena is not as great as the unity of either physical or chemical phenomena, and so on. All of this provides the relevant ontologic foundation for singling out regional concrete-science interdisciplinarities. According to the "condensation" and the specific type of unity of this or that sphere of reality studied by some region of the sciences, there must necessarily exist a corresponding "condensation", consolidation and specification of interdisciplinarity.[5]

Within the sphere of regional concrete-science interdisciplinarity there will be found a number of specific philosophical and methodological problems, the main of which is the quest for the most fundamental unity of the whole aggregate of disciplines in the given region. This involves the reductionist approach,[6] whose substance is to explain the complex by reducing it to the simple, the superior, by reducing it to the inferior. It exists both in the disciplinary and in the interdisciplinary form. The interdisciplinary form of reductionism consists in an explanation of complex phenomena studied by one discipline in the light of simpler phenomena studied by another discipline. In the event of regional interdisciplinarity connected with the quest for the most fundamental unity of the whole region, reductionism amounts to an effort theoretically to substantiate and explain phenomena in the given sphere of reality studied by different disciplines of the region in the light of the principles and laws of one of these disciplines which proves to be the most fundamental.

There are substantial grounds for regional reductionist interdisciplinarity. The point is that the most fundamental phenomena and processes of some area of reality are included in its other phenomena and processes as component parts. But the specifics of the parts do indeed determine (even if partially) the specifics of the whole. However, as the general systems theory, cybernetics, biology and even physics and chemistry demonstrate ever more fundamentally, the specifics of the whole do not in any sense amount to a sum of the specifics of its parts. The latter determine the properties of the whole partially, instead of entirely. Because the whole is always more than the sum of its parts, reductionism is always less than and falls short of the ultimate cognition of the whole.

The systematization of all the interdisciplinary processes specific for the given region of the sciences and the definition of the tendencies in their further development are another important problem in the study of concrete-science interdisciplinarity. This kind of systematic exploration of the existing and potential interdisciplinarity of the given region can alone lead it out of the chaotic empirical state in which we now find it.

4.1.2 Borderland Interdisciplinarity

The so-called borderland, or transitional, allied, adjacent sciences are among the most developed forms of essentially ontologic interdisciplinarity. The objects of these sciences are formed at the "junction" between the superior and inferior levels in the motion of matter. They study the transitions from physical objects to chemical and biological objects, and from biological to social objects. They also study the transitions from inferior entities to superior ones within the boundaries of some single level of movement.[7]

There are now dozens of such sciences. Back in the 19th century, there-emerged such borderland interdisciplinarities as astrophysics, physical chemis try, geophysics, psychophysics, etc. Throughout the first half of the 20th century there was fairly intensive formation of such sciences. We now find at the stage of formation social psychology[8] and sociobiology, although there is heated discussion with respect to the latter.[9]

The central problem facing such sciences which has yet to be given a sufficiently complete philosophical solution is to define the specifics of the subject-matter of the borderland interdisciplinarity.[10] To this very day, some scientists say that the subject-matter of biophysics, for instance, consists of biological phenomena, but studied by means of physical methods,[11] whereas others insist that its subject-matter consists of physical processes but existing in a special biological form.[12] There appears to be more ground, after all, to assume that the subject-matter of any borderland interdisciplinarity always comes to a study of manifestations of the inferior in a superior mode of existence, that is, specific manifestations of the inferior which can exist only when included within the composition of the superior, and not in some form which is free and independent of the superior. This means that biophysics, for instance, is the physics of life, and not life studied by physical methods. In the phenomena of life, physical methods can be used to study only physical phenomena, even if these are specific and have been transformed by the phenomena of life.

Another important philosophico-methodological problem of borderland interdisciplinarity is to comprehend the specifics of the reductionism with which it is connected. In contrast to regional interdisciplinary reductionism, borderland reductionism amounts to the reducibility of the superior to the inferior not within the framework of a homogeneous region of disciplines but between the frameworks of such regions. Such reductionism is, within definite limits, just as successful as regional fundamental reductionism. It actually helps to derive, for instance, definite biological phenomena from physical and chemical regularities.[13] But it, too, has turned out to be not absolute and complete, but relative and partial. It does not bear on all the phenomena of the superior sphere, but only on a definite range of these.

Thus, quantum chemistry, being an explanation and prediction of chemical phenomena on the basis of the principles and theoretical structures of quantum mechanics and electrodynamics, requires the introduction of new propositions which do not follow from the postulates of quantum physics to explain even the simplest chemical system, namely, the hydrogen molecule. Many data in modern molecular biology also testify to the impossibility of fully deriving the specifics of microbiological processes from physical and chemical regularities. An important indicator of the impossibility of complete reduction of molecular biology to physics and chemistry is provided, in particular, by the operon theory of gene. According to K. F. Schaffner, "it would seem that thought the operon theory was articulated at a level which is very close to the physico-chemical level, it *itself* is not physico-chemical theory. The operon theory was constructed, formulated, and tested in terms and using the techniques of neo-classical genetics, a science which does not *require* a reduction to physics and chemistry".[14]

4.1.3 System-Transborder Interdisciplinarity

In the event of borderland interdisciplinarity we have to deal with objective processes and entities which are a common field of two allied disciplines because the object of one of these in effect falls within the object of the other, constituting a specific cross-section, a layer of the inferior within the superior. At the same time, the ontological unity of various objects of the world also appears in the form of processes and systems which are common either to some entire sphere of phenomena studied by the corresponding region of sciences or to several or all of these spheres. Thus, the ontologic unity of the objects of social sciences largely assumes the form of specific systems and processes in social life which, in one way or another, whether immediately or otherwise, fall within the objects of all or a sizable group of social sciences. Among such processes are, for instance, aim-pursuing human interaction, learning, perception and processing of information, organization and control of one or another human activity, etc.[15] Nowadays, such processes have already become the object of the corresponding theories, like theory of games, learning theory, information theory, organization theory, cybernetics, and so on. These scientific formations have inevitably to fulfil a highly extensive and specific interdisciplinary function.

The specific feature of this form of interdisciplinarity consists in the fact that here the mechanism of science integration is interaction not only of allied disciplines but also of those which are located far from each other within the overall hierarchy of the system of sciences. The interaction of natural, technical and social sciences (global transborder interdisciplinarity) or of all the social, or of all the technical, or of all the biological, and so on, sciences (regional transborder interdisciplinarity) most frequently occur when such synthetic, complex sciences and theories take shape.

The object of transborder interdisciplinarity is always a specific system (or process) which in its abstract form turns out to be common to many concrete spheres of reality, and exists in these spheres in the form of manifestations specific in their material nature. Consequently, this object is one or another of kinds of abstract, universal systems and processes. Cybernetics, for instance, makes a study of abstract, universal control systems. Its basic task is to disclose the objectively existing, as well as capable (in virtue of the corresponding laws of the world) to exist, abstract structures of control, which are isomorphic with respect to the concrete material nature of the various biological, technical or social control systems.

In design and task, general systems theory is the broadest possible one of the concrete transborderline interdisciplinarities. It is based, L. Bertalanffy emphasises, on the fact that "the world, i.e., the total of observable events, shows structural uniformities, manifesting themselves by isomorphic traces of order in the various levels or realms."[16]

4.1.4 System-Complex Interdisciplinarity

Alongside transborder systems and processes, the ontologic unity of the world is also expressed in the presence in its various spheres of entities and processes that constitute a single, integral and interconnected complex of qualitatively distinct subdivisions, which are such that each of these can be the object of a corresponding scientific discipline or even of a group of disciplines. The existence of transborder type of systems and processes, as has been shown, is determined by links and law-governed relations among many or all spheres of reality which are universal and isomorphic in the sense of their being independent of the qualitative specifics of these spheres. The existence of the system-complex formations and processes under consideration is determined by another type of objective links and regularities. The specifics of these links and regularities consist in the fact that they determine a conjunction and interaction of diverse departments, parts, and elements, as a result of which some sufficiently global inseparable-integral entity is formed. Consequently, these are special systems-forming and systems-organising links and regularities. Their operation results in that the changes occurring with such system-complex entities are governed not so much by general universal physical, chemical, biological and other laws, as by modified manifestations of these laws, while the modifications themselves are determined by the systems integrity of the given entity, by the systems links prevailing within it.

A study of any system-complex entity as a rule starts with the comprehension of its individual parts and large departments by corresponding independent sciences. Here, the systems-forming relations between these departments as a rule remain in the background. Initially, the system-complex entity does not appear in the cognition as such, as the systemic. Its individual departments are analysed not as its departments, but as absolutely autonomous formations. Thus, for a long time science made a study through its corresponding disciplines not of the biosphere, but of its individual and independent departments. Only today, the complex of sciences of the biosphere is beginning to reveal systems-forming links between its individual departments.[17] Similarly, for a long time linguistics studied language phenomena simply as such, altogether leaving in the background the social and psychological aspects of language and its objective systems character, which includes not only the purely linguistic but also psychological and social processes.[18]

However, sooner or later, the objective system-complex nature of this or that entity is inevitably discovered by science or by the sciences studying its individual departments. In such cases, there begins a specific process or complex of processes whose general orientation is a theoretical re-establishment of the objective integrity and systems nature of the complex entity which has evaporated or which has not yet been discovered in the course of earlier theoretical studies of the given entity. This process constitutes a peculiar—system-complex—form of interdisciplinarity.

This form acquires ever greater specific gravity among the ever more extensive processes of interdisciplinary development in modern science. A sizable number of present-day interdisciplinary scientific formations are nothing but system-complex sciences (if not actually, then at any rate potentially: both in terms of objective grounds and the overall tendency of their development).

As for the object of such interdisciplinarity, in general form it is the specific, systems-forming and systems-organizing links and regularities thanks to which the complex of some subdivisions of reality appears as an integral system. Only as a result of the discovery of these regularities can a complex of some disciplines elaborate a common theoretical structure to link up these disciplines in a single systems interdisciplinary entity. It goes without saying that this process leading to the formation of system-complex interdisciplinarity is extremely complicated and labour-intensive. In effect, to this day there is not a single sufficiently mature and fully formed system-complex interdisciplinarity. All of these are at the stage of more or less successful formation. It appears that the science of science has made the greatest headway in its formation as a system-complex interdisciplinarity.[19]

4.2 EPISTEMOLOGICAL FORMS

Epistemological interdisciplinarity, while being objectively based on this or that ontological unity of the objects of different disciplines, operates with relative independence from the ontological forms of interdisciplinarity, although it does reach its state of flowering only when the latter are correspondingly developed. What is more, as an aggregation of "soft" structures of science, which are relatively independent of its "rigid" structures, epistemological interdisciplinarity emerges much more easily than ontological and sociological interdisciplinarity. It may link up some disciplines without waiting for the level of their development to enable them to reflect the objective community of their subjects-matter. In principle, such a "soft" link between disciplines, having barely clung to perhaps a very weak comprehension of the "rigid" unity of their subjects-matter, leads to a disclosure of profound objective links and regularities of the unity of these subjects-matter. The "soft" structures of one science, being more developed than those of another, are frequently capable of pulling up to the surface of scientific analysis, by means of empirical material from that other science, the "rigid" structures of both sciences to the extent to which they objectively coincide with each other. Therein lies the tremendous heuristic role of the epistemological interdisciplinary interaction of the sciences. Strictly speaking, interdisciplinary development of modern science is most frequently embodied in epistemological forms rather than being really based on an already recognized ontological unity of disciplines.

In accordance with the general epistemological structure of science there is a need to single out within the system of the epistemological forms of interdisciplinarity its following types, which are more or less mature and frequently occur in real interdisciplinary processes: interaction of the conceptual apparatuses of the sciences, interaction of scientific disciplines by means of their laws and the structures of their laws, interaction of disciplinary theoretical structures, and interaction of the methods of individual disciplines.

These types constitute the main levels of the epistemological interaction of the sciences. In principle, this interaction may start at any of these levels, although most frequently it begins either at the conceptual or the methodological level. But at whatever level it may start, it cannot always remain only on one of these. Sooner or later, it is bound to include within itself the whole system of these levels, thereby being converted into a concrete but systems form of epistemological interdisciplinarity which is complete from the standpoint of all these levels.

For lack of space, below I give only the most general definitions of the concrete epistemological forms.

In the first place, we must take into consideration the specific processes of development of the epistemological structures related to the above-mentioned ontologic forms of interdisciplinarity.

Regional Foundationalization of Disciplines. It is a process in the course of which the "soft" structures of the most foundational, basic discipline of the region penetrate all the other sciences of the given region. This is, therefore, epistemological reduction of the disciplines of the given region to the epistemostructures of its most basic discipline. For present-day physics, for instance, it is the "microphysicalization" of the whole of physical knowledge.

Regional interaction of epistemostructures. It is a process of the mutual development of the "soft" structures of all the disciplines of the region. It results in the corresponding systems of concepts, theoretical structures and efforts which are common to the whole given region.

The foundationalization of borderland interdisciplinarities. It is the influence of the "soft" structures of the inferior and more basic, foundational from two bordering, adjacent sciences on such structures in the superior and less basic science. In the case of biophysics, for instance, it is the physicalization of its "softness".

Interaction of epistemostructures in borderlands. It is not only foundationalization, that is, the transformation of the "soft" structures of borderland science on the basis of the "softness" of its inferior component, but also the influence of the superior "softness" on the inferior. For biophysics, or biochemistry this is not only its physicalization or chemicalization but also its biologization.[20]

Epistemological system-transborder interdisciplinarity. It is a process of the formation of specific epistemostructures of system-transborderline sciences. For instance, the formation of concepts, laws, principles, theoretical structures and methods of cybernetics.

Epistemological system-complex interdisciplinarity. It is the shaping of specific epistemostructures making it possible to realise on a theoretical level the objective system-complex integrity of some subdivisions of reality which had earlier appeared only as separate objects of the independent disciplines.

In the second place, investigating epistemological forms we are bound to take into account the transborder epistemological processes. They are the various concrete processes in the development of the "soft" structures of all the sciences or of the greater part of them. The most important of these processes at the present time are the following:

Mathematization, which is the application in this or that science not only of the mathematical structures elaborated in mathematics itself or in other essentially mathematized disciplines, but also the elaboration of specialized mathematical structures by the given science itself. Modern mathematization largely becomes more than a mere borrowing of mathematics by the given science, and includes the creation by that science itself of fundamentally new mathematical structures which are more adequate to the specifics of its object than the borrowed structures.

Physicalization, which is the influence of physical epistemostructures on the epistemological equipment of all the other sciences. In modern science, physicalization must become not only an application of universal physical ideas and methods in other sciences, but also the elaboration by the given science itself of its specific physical aspect, its comprehension of the physics of its own object, and the elaboration of methods precisely for that object. Otherwise, physicalization, like mathematization, incidentally, threatens to become a vulgarization of the specifics of the given science or to boil down to meaningless trivialities.

Biologization. It is the influence of the epistemostructures elaborated by modern biology on the epistemostructures of other sciences.

Cybernetization. It is the application of the ideas and methods of cybernetics in other sciences.[21]

Informatization. It is the application of concepts, theoretical structures and methods of the theory of information in other fields of science.[22]

Logization and formalization.[23] It is the application of the structures of logic, especially mathematical logic, in other sciences, the conjunction of formalization of logic and formalization of the concrete content of this or that science and the transfer of formalized languages from one field of science to others.

Automatization of scientific cognition.[24] It is the elaboration of methods for staging experiments with the use of automatic technical systems, and also the methods for processing the results of experiments with the use of computers.

Modelization of science.[25] It is the application in various sciences of the method of modelling and development in this context of relevant methods of specific model epistemological conceptions.

Systems analytization.[26] It is the use in the sciences of systems-structural, structure-functional methods, systems analysis, and elaboration of the required epistemological problems.

General conceptualization and unification of scientific language.[27] It is the formation of concepts and the creation of a language common to all the sciences. The main thing in this process is unification of the language of science as an expression of the objective processes of interconnection between various branches of scientific knowledge. Here, the common language of science can be worked out only as a result of interaction among all the scientific disciplines, and not through a reduction of the whole scientific language to the language of some single science.

Integration of the empirical and the theoretical.[28] It is the formation in all the sciences of ever closer unity of empirical and theoretical levels of cognition, and also mutual exchange by the sciences of empirical and theoretical material: use of empirical material provided by one science to elaborate the theoretical structures of another, and the use of the theory of one discipline for processing the empirical material of another.

The important place among epistemological forms is occupied by the most extensive means of epistemological interaction of sciences.

These forms of interdisciplinarity encompass (in one way or another) as their own elements all the regions and areas of scientific cognition, all the disciplines. These forms establish various mutual links among the scientific disciplines, thereby inevitably promoting to a greater or lesser degree the entry of these disciplines into closer interaction with each other.

Classification of sciences. Considering that classification of the sciences is the establishment of a definite connection, of an hierarchical coordination and subordination of scientific regions and spheres which are part of these regions, and disciplines relating to the various regions and spheres, it inevitably operates as the special broadest epistemoform of interdisciplinarity. It is a constantly developing interdisciplinary entity. At every stage of its development, it reflects the degree attained in the scientific comprehension of the breakdown of reality (as well as the cognition of reality) into separate subdivisions, and also realization of the links between the individual subdivisions of reality as well as of science. Considering that the classification of the sciences has important practical functions to perform, including functions in the sphere of organisation of the structure of scientific institutions and interrelations between them, its constant improvement is among the most important problems in the further development of interdisciplinarity.

Concrete-Scientific Picture of the World.[29] Such the broadest form of interdisciplinarity as a scientific picture of the world holds the peculiar place among the varied processes of interdisciplinary development of modern science. Its structure and functions, as a specific interdisciplinary entity, have not been adequately studied. However, it is quite probable that it now constitutes the most substantial form of the broadest front of interaction among the sciences. In the most general terms, it emerges as some interdisciplinary theoretical system generalizing the most essential empirical data and the most important theoretical constructions of all the regions and areas of scientific cognition, and of all the scientific disciplines in principle. However, it is a multiple, hierarchical entity. The first thing that stands out within the structure of the concrete-science picture of the world is the physical picture of the world, its most fundamental and theoretically elaborated level. On top of this, superstructured and interacting with it are the general picture of biological processes and the general picture of social processes. Here, the regional pictures of the world always have a tendency, as physics and biology show in particular, to grow into elaborate regional theories (general theoretical physics, general theoretical biology).

For lack of space I cannot go into any detailed analysis of the structure and functions of the concrete-science picture of the world, but I must mention at least two of its key interdisciplinary functions. First, the concrete-science picture of the world helps the initial elaboration of theories in disciplines which are dominated by the empirical level in the comprehension of reality. For instance, geology structures its primary theory of geological processes by making use of the physical picture of the world. Second, the concrete-science picture of the world creates the necessary conceptual climate for transferring various epistemostructures and above all methods from one field of science to others.

Concrete-Scientific Style of Cognition.[30] The scientific picture of the world is an interdisciplinary entity connecting various disciplines and above all the major regions of the sciences through a generalization of their content.

It is a contental (non-formal, contrary to formal) interdisciplinary amalgamation, the amalgamation of sciences from the standpoint of what is being cognised, from the standpoint of the chief results of the scientific mastering of the world. An amalgamation of the sciences from the standpoint of how cognition is effected, by means of which methods, approaches, logical instruments and forms results in an interdisciplinary entity like the concrete-sciencific style of thinking in the given historical epoch. This type of the broadest concrete-methodological amalgamation of the sciences contains a generalization of the most important methodological advances in physics, chemistry, biology, the social sciences, the sciences of man, and mathematics.

Philosophical Picture of the World. This picture is the philosophical world outlook of the given historical epoch. It is the broadest possible interdisciplinary entity of the sciences in the event that it is a philosophical generalization of the concrete-science picture of the world, a comprehension of the most important advances in the concrete sciences through the prism of philosophical categories. Of course, not every philosophy will be so bold as to produce a generalization of the advances in the concrete sciences. Discussions at the 15th World Philosophical Congress in Varna in September 1973 showed that some philosophers contended that philosophy is not a science at all, that science has no need for any philosophy.[31] Of course, such philosophers will not and cannot produce a philosophical picture of the world as the broadest possible interdisciplinary entity. But it is produced by other philosophers who insist on the intimate link between philosophy and science.

Philosophical Methodology of Science. Comprehension of the methodological advances in the concrete sciences through the prism of philosophical categories and methodological advances in philosophy itself yields a specific interdisciplinary methodological entity. It is an interdisciplinary entity because it operates more or less effectively in the concrete sciences, helping them to solve their own methodological problems. The philosophical methodology of science acts as the most global methodological interdisciplinarity, just as the philosophical picture of the world gives the most global contental amalgamation of the sciences.

4.3 SOCIO-ORGANIZATIONAL FORMS

The interdisciplinary development of the sciences can also proceed without the formation of relevant scientific establishments or any complex scientific communities. However, this or that form of interdisciplinarity becomes truly mature and a really fruitful process of interaction among the sciences only when it is invested with socio-organizational forms that are adequate to its tasks and potentialities. Within the overall system of forms of interdisciplinarity socio-organizational ones have the peculiar functions of realizing in practice the potentialities for extending and deepening the scientific getting hold of reality which are latent in this or that concrete interdisciplinary formation.

Below I have to confine myself to a few brief remarks on the most important socio-organizational forms.

First of all, it might be well to point out such the oldest and simplest method of socio-organizational cooperation among the disciplines as an exchange of resources for researches (exchange of scientific personnel, scientific-experimental equipment, etc.).

Next are the forms of planning and organization of researches: coordination of the research activity in different sciences, organization of the complex research projects, working up of the unified scientific-technological policy at the national or regional levels, international scientific-technological policy as the most global and broadest socio-organizational form of interdisciplinarity.

The growing ties between science, technology, material production, social progress and the development of the human personality and individuality are becoming a most characteristic feature of contemporary development of science. This has produced new socio-organizational forms of integration of science and the life of society: science-production associations, applied and even basic research oriented upon practice, and so on. These forms are inevitably interdisciplinary, and include a specifically interdisciplinary aspect. This springs from the impossibility of determining the social and ecological consequences of this or that scientific or technical achievement only on the strength of the knowledge worked out by the scientific discipline effecting the achievement. It is now quite obvious that the practical application of science should proceed not only from the content of this or that scientific accomplishment but also from its immediate and remote social and ecological consequences. It is just as obvious that without the involvement of the most diverse disciplines neither the full scope nor the correct form of any, whether immediate or the most remote social consequences of the application of the achievements in this or that discipline can be determined. It is for this reason that of late much attention has been given to the interdisciplinary analysis of the social consequences of modern scientific and technological progress. As the UNESCO Conference on the Scientific and Technological Revolution and the Social Sciences (Prague, September 6—10, 1976) noted, "UNESCO tries to establish close collaboration between the social and natural sciences to improve the understanding and analysis of the social aspect of the scientific and technological revolution." The state of things here is such that one could risk prognosticating the emergence in the near future of special interdisciplinary departments of science whose main function will be an evaluation of the immediate and longterm perspectives of the social influence exerted by the various branches of scientific and technical progress. For the time being, the essentially interdisciplinary evaluation of the social consequences arising from the application of science and technology has gone forward in forms which are organically connected with the forms of the practical application of science, which have either already taken shape or are in the process of formation. It is also expressed in the form of some specific socio-organisational processes within the body of science, which are common to all science in scope. I have merely remarked on these processes, without giving any detailed description, for lack of space.

Technization and industrialization of researches is the interdisciplinary development of science through its growing links with technology in the sphere of scientific research itself and the transformation of the production of scientific knowledge into a process which is industrial in scope and mode. The technization and industrialization of knowledge also proceed as a specific interdisciplinary socio-organizational process. Besides, they go to create a totally new material and organizational basis for a form of interdisciplinarity like complex (combined) research.

Cosmization of science[32] is interdisciplinary development of science connected with theoretical and practical getting hold of cosmic space. More and more sciences are at present launched out on a voyage of cosmic orbits, adventures and discoveries. And here they more and more tightly interact with each other in a form of interdisciplinary by their essence concrete cosmic research projects. The scope of these projects encompasses both the most advanced of science like physics, chemistry, mechanics, astronomy, geology, meteorology, biology, psychology, law, economy and others. Moreover, it is exceedingly important that cosmic researches are carried on not only as complex, but also as large interdisciplinary and now more and more often international projects.

Ecologization of scientific enterprises is inclusion within both the theoretical and practical projects of this or that science of essentially interdisciplinary aspects connected with the need to determine the impact of these projects on the state of this or that biospheric environment or of the biosphere as a whole.

Sociologization and axiologization of science is inclusion of social and axiological aspects within any of the scientific disciplines. Any modern science now begins to develop and be practically applied not only in the light of its own needs, but also of certain social requirements and values, of the various social consequences of its development and application. Even in the definition of new lines of basic research, which had earlier, as a rule, been derived from the internal logic of the development of the science itself, the question now ever more frequently and inevitably arise whether some research can be started only on the strength of the fact that it springs from the internal logic of the development of the science, or whether its results could lead to substantial undesirable social consequences. With respect to modern genetics, for instance, many scientists believe that it is already undesirable to adopt an unconditional claim to absolute freedom in the quest for scientific truth.[33]

This single process of the sociologization and axiologization of scientific cognition inevitably unites the individual disciplines and operates as a powerful factor in the integration of the natural and social sciences. In order to bring out their social and value aspects, the sciences must inevitably exchange information and to work out some common criteria for the social and value consequences of their development.

The humanization of science[34] is the pivoting of the whole of science on the problem of man, the adoption as the main line of development both in the social and in the natural and technical sciences their increasing orientation upon complex (combined) researches into man, upon theoretical comprehension of the whole of man's multi-aspect (biological, social, emotional and intellectual) integrity and the practical development of every aspect of this integrity. The human aspects of science are increasingly established as the real centre of all the theoretical and practical aspirations of science round which the whole global process of its development and application is crystallized, ranging from the most intricate and abstract theoretical constructions to the simplest and most concrete practical applications.

That being so, the centre of all the aspirations of science, including the interdisciplinary aspirations, being found, it remains nothing to do but to complete this survey of the basic forms of interdisciplinary development of modern science. If anything of importance in these forms has been missed, it still has to be contained in this centre of interdisciplinarity, so that in principle it could be derived from it by means of the relevant logical reasoning, however protracted and difficult this may be.

REFERENCES

[1] Some important interdisciplinary relations between modern sciences are described in the following essays:
Physical theory (*Philosophical-methodological analysis*), Moscow. Nauka, 1980; (in Russian); Lyapunov A. A., *Problems of theoretical and applied cybernetics*, Moscow, Nauka, 1980 (in Russian); Mirsky E. M., *Interdisciplinary researches and disciplinary organization of science*, Moscow, Nauka, 1980 (in Russian); Volkenshtein M. V., *Physics and biology*, Moscow, Nauka, 1980 (in Russian); Ursul A. D., "The union of philosophy and natural sciences", — *Voprossi filosofii*. Moscow, 1980, No 6, p. 46—61 (in Russian); "Anthropologie, sciences humaines, sciences naturelles", — *L'anthropologie en France*, Paris, 1979, p. 25—263; Leary D. E., "Wundt and after: Psychology's shifting relations with the natural sciences, social sciences, and philosophy", — *Journal of the history of the behavioral sciences*, Brandon, 1979, vol. 15, No 3, p. 231—241; Machlup F., *Methodology of economics and other social sciences*, N. Y. etc., Acad. press, 1978; Neill W. T., *Archeology and the science of man*, N. Y., Columbia univ. press, 1978; Mauss M., *Sociology and psychology*, London etc., Routledge a Kegan Paul, 1979; *Brain, environment, and social psychology*, Baltimore, 1979; *The humanities in a computerized world*, Albany (N. Y.), 1979; Snyder P., *Toward one science: the covergence of tradition*, N. Y., 1978; Crutchfield M. A.,

Elementary social studies: an interdisciplinary approach, Columbus etc., Merrill, 1978; Hall G. M., *Farewell to Darwin: the unified field theory of physics, the genetic process, and psychology*, St. Louis (Mo.), Green, 1977; Becker E., *The structure of evil: an essay of the unification of the sciences of man*. N. Y. 1976.

2 See for example: Albert J., Herlitzius E., Jobst E., "Probleme der Einheit von Natur-, Technik- und Gesellschaftswissenschaften", — *Wiss. Ztschr. der Techn. Univ. Dresden*, 1980, Jg. 29, H. 2, S. 281—292; Smith N., "Symptomatic silence in Althusser: the concept of nature and the unity of science",—*Science and society*, N. Y., 1980, vol. 44, No 1, p. 58—81; Holz H. H., "On the unity of the natural and social sciences",—*Revolutionary world*, Amsterdam, 1979, vol. 34—35, p. 60—73; Heller A., "Can the unity of sciences be considered as the norm of sciences?"—*Counter-movements in the sciences*, Dordrecht, etc., 1979; Causey R. L., *Unity of science*, Dordrecht-Boston, 1977.

3 See: Kneller G. F., *Science as a human endeavor*, N. Y., Columbia univ. press, 1978; Mendelson E., Weingart P., Whitely R., *The social production of scientific knowledge*, Dordrecht-Boston, 1977.

4 *The structure and development of science*, Dordrecht-Boston, 1979; Chalmers A. F., *What is this thing called science? An assessment of the nature and status of science and its methods*, St. Lucia, Univ. of Queenslandpress, 1978.

5 On problems of some regional concrete-science interdisiplinarities see, for example: Weizsäcker C. F. von, "The preconditions of experience and the unity of physics", —*Transcendental arguments and science: essays in epistemology*, Dordrecht etc.- 1979, p. 123—158; Bohm D., "The implicate order: a new order for physics". — *Process studies*, Claremont, 1978, vol. 8, No 2, p. 73—102; Matumana H. R., Varela F. J., *Autopoiesis and cognition: the realization of the living*, Dordrecht etc., Reidel, 1980; Warren Ch. E., Allen M., Haefner J. W., "Conceptual frameworks and the philosophical foundations of general living systems theory",—*Behavioral science*, Baltimore-Ann Arbor, 1979, vol. 24, No 5, p. 296—310; Giesen B., "Gesellschaftliche Identität und Evolution",—*Soziale Welt*, Göttingen, 1980, Jg. 31, H. 3, S. 311—332; Gray W., Shay D., "Defining social science: who can join the club?"—*Social science*. Winfieid, 1979, vol. 54, No 4, p. 231—235.

6 *Materie, Leben, Geist: zum Problem der Reduktion der Wiss. B.*, Duncker a Humblot, 1979; Čížek F., "Biologie, fyzika a redukcionismus",—*Filoz. časpis*, Praha 1979, roč. 27, č. 4, s. 488—503.

7 See: Jones D., "The boundaries of chemistry",— *New scientist*, London, 1979, vol. 83, No 1170, p. 661—663.

8 *Emerging strategies in social psychological research*, Chichester etc., Wiley, 1979; *Introducing social psychology*, Hardsmodsworth, Penguin books, 1978.

9 Bogdany F. J., "Soziobiologie — Möglichkeiten und Grenzen der neuen Synthesis", — *Kölner Z. für Soziologie und Sozialpsychologie*, 1980, Jg. 32, No 2, S. 312—324; Baldwin J. D., Baldwin J. I., "Sociobiology of balanced biosocial theory?" — *Pacific sociological review*, San Diego, 1980, vol. 23, No 1, p. 3—27; *The sociobiology debate*, N. Y. etc., Harper a Row, 1978; Christensen Y., *L'heure de la sociobiologie*, Paris, Michel, 1979.

10 *The nature and status of ethnobotany*, Ann Arbor, 1978; Godelier M., "The object and method of economic anthropology"—*Relations of production: marxist approaches to economical anthropology*, London 1978, p. 49—126; Boutillier R. G. et al., "Crises in the two social psychologies: a critical comparison",—*Social psychology*, Albany, 1980, vol. 43, No 1, p. 5—17.

11 Kedrov B. M., "The dialectical way of the theoretical synthesis of modern natural

science knowledge". — *The synthesis of modern scientific knowledge*, Moscow, Nauka, 1973, p. 28 (in Russian).

12 *Methodological and theoretical problems of biophysics*, Moscow, Nauka, 1979, p. 7.

13 Vollmer G., "Einheit der Natur und Einheit der Wissenschaft — Probleme des Reduktionismus für die Biologie", — *Wittgenstein, der Wiener Kreis und der kritische Rationalismus*, Wien, 1979, S. 318—324.

14 Schaffner K. F., "The unity of science and theory construction in molecular biology", — *Philosophical foundation of science*, Dordrecht-Boston, 1974, p. 513.

15 See: Apostel L., "Interdisciplinary relations among the sciences of man", Preliminary report, UNESCO, 1976.

16 Bertalanffy L. von, *General systems theory*, N. Y., 1968, p. 49.

17 "Conceptual issues in ecology", — *Synthese*, Dordrecht, 1980, vol. 43, No 1, p. 1—181, No 2, p. 195—342; Dunlap R. E., Catton W. R., "Environmental sociology: a framework for analysis", — *Progress in resource management and environmental planning*, Chichester etc., 1979, vol. 1, p. 27—85; *Environmental knowing: theories, research and methods*, Stroudsberg, 1976.

18 Rabanales A., "Les interdisciplines linquistiques", — *Linguistique*, Paris, 1979, vol. 15, fasc. 2, p. 95—105.

19 *Perspectives in metascience*, Göteborg, 1979.

20 Lane B. G., "The language of biochemistry", — *Canadian research*, Toronto, 1979, vol. 12, No 4, p. 31—34.

21 *Cybernetics and modern scientific cognition*, Moscow, Nauka, 1976.

22 Ursul A. D., *The problem of information in modern science*, Moscow, Nauka, 1975 (in Russian); Davis W. S., McCormack A., *The information age*, Reading (Mass.) etc., Addison — Wesley, 1979; *Advances in information systems science*, N. Y. — L., 1978.

23 *Physical theory as logico-operational structure*, Dordrecht etc., 1976; Taylor Ch., "Formal theory in social sciences", — *Inquiry*, Oslo, 1980, vol. 23, No 2, p. 139—144; *Logical problems of the research of scientific cognition*, Moscow, Nauka, 1980 (in Russian).

24 See, for example: Coats R. B., Parkin A., *Computer models in the social sciences*, London 1977.

25 Destouches J.—L., "La modélisation et la sémantique", — *La sémantique dans les sciences*, Paris 1978, p. 173—179; *Modèles et interprétations*, Lille 1978.

26 Cavallo R. E., *The role of systems methodology in social science research*, Boston etc., Nijhoff, 1979; Klir G. J., "The general systems research movement", — *Systems models for decision-making*, Bangkok, 1978.

27 *Theoretical and methodological problems of terminology*, Moscow, International symposium, 27—30 Nov. 1979; Febler H., "International standardization of terminology: theoretical and methodological aspects", — *International journal of the sociology of language*, 1980, No 23, p. 65—79.

28 White D. H. etc., "The interdependance of theory and experiment in revolutionary science: the case of parity violation". — *Social studies of science*, London 1979, vol. 9, No 3, p. 303—327; *Empirie, teorie a praxe ve společenských vědách*, Praha 1979.

29 Yulai U. V., "Epistemological aspects of the notion Scientific picture of the world" — *Philosophical problems of modern natural sciences*, Kiev, 1980, vol. 49, p. 31—37 (in Russian).

30 Serebryakova E. S., *Styles of cognition as the landmarks in the scientific cognition*, Dissertation thesis, Moscow, 1980 (in Russian).

31 Mercier A., "La philosophie et la science", — *Proc. of the XVth World Congress on Philosophy*, vol. 1. Sofia, 1973, p. 25.

[32] See, for example: Ursul A. D., *Mankind, Earth, Universe,* Moscow, Mysl, 1977; *The impact of space science on mankind,* N. Y.—L., 1976.

[33] Motilski A. J., "Brave new world? Current approaches to prevention, treatment and research of genetic diseases raise ethical issues".—*Science,* vol. 185, No 23, 1974, p. 136.

[34] *Science, technology and the human prospect,* N. Y. etc., 1980; Cantore E., *The humanistic significance of science,* N. Y., 1977.

5. Remarks on the Dialectical-Materialist Type of Rationality

JINDŘICH ZELENÝ

A great creative process is going on in our epoch: the transition from bourgeois to communist forms of social and human life. Our point of departure is that part of this world process is the rise, development and progressing social application of *a new historical type of scientific rationality*, and we shall offer some characteristics of this, especially in comparison and partly in contrast with the Cartesian type of scientific rationality.

First, of course, we must outline what we mean by "type of (scientific) rationality".

In every epoch, and therefore in ours as well, both theoretical and ordinary thought are historical products, linked not only to each other and to language, but also to historically changing forms of human work and life. The term "rationality" is for us an abstraction from the process of human cognitive thinking and conscious action[1] and is meant to encompass the logical-ontological foundations of thought. It applies both to theoretical (i.e. scientific in a broader sense) and to everyday, ordinary thought.

In spite of the great diversity and varying depth of changes in the development of both theoretical and ordinary thought (and in their interconnection, too), we think it possible — merely by concentrating on the logical-ontological foundations of thought — to distinguish some basic or pre-eminent historical types of rationality. A type of rationality can be characterized by applying such joint criteria as e.g.: What are the basic categories and methods of thought under consideration? What conception of the relationship between theory and practice and between descriptive and evaluative statements has been displayed? In other words, different types of rationality exhibit variations in their conception of logical-ontological problems which have several levels and can be roughly classified into a few paragraphs dealing with: 1) categories;

and a more or less elaborated theory of categories; 2) predication and a more or less elaborated theory of predication; 3) modes of reasoning and rational argument, i.e. what we could perhaps call — by combining a Hegelian and a Fregean term — "Forminhalte der Gedankengefüge"; 4) the problem of truth which includes fundamental logical-ontological questions concerning the relationship between the logical and the historical, between social forms of *praxis* and forms of thought, between natural forms and reflecting forms of thought mediated by actual human *praxis*.

The distinction between different historical types of rationality is not absolute, not relativistic, not without continuity. Historical types of rationality exhibit not only distinctions and differences but also include many important common invariations.

If we limit our view only to scientific rationality, at least three pre-eminent[2] types of it can be distinguished in our cultural tradition — the Aristotelian, the Cartesian and the Marxist-Leninist (the dialectical-materialist). This is not to say that these are the only three main rationality types or that they cover the whole history of scientific and theoretical culture. The suggested selection of three pre-eminent types of scientific rationality seems appropriate and fruitful to provide relevant comparative material for deepening our insight into contemporary changes in the logical-ontological foundations of scientific thought.

Focussing now on the question of what main differences there are between the Cartesian and the modern dialectical-materialist type of scientific rationality, we can point out the historizing of forms of being and thought as the first main distinctive feature of the latter. While the former may be called *ahistorically technical*, the latter is *historical* and *practical*. However, such general characteristics as "historizing" do not tell us much if they are not qualified and specified. Let us try to do just this — by confronting the dialectical-materialist critique of the Cartesian framework with other critiques of it, which in the 20th century, are rather frequent.

The most significant specification may perhaps be expressed by saying that the historizing of forms of being and thought and the practical foundation of thought within the dialectical-materialist type of rationality does not deprive cognitive thought of *objectivity*. On the contrary, it makes cognitive thought more and deeply *objective*[3] than was cognitive thought within the Cartesian framework.

This is the point in which Heidegger's and the dialectical-materialist critiques of the Cartesian framework principally differ. According to Heidegger, all modern scientific thought that originated in the 17th century and developed conjointly with technological application in machinery, has proceeded on the basis and within the limits of the Cartesian metaphysical framework. To the question of which metaphysical basis, what conception of being and truth is implied in modern science, Heidegger answers: "Research masters reality if it can either precalculate its future course or ex post calculate its past course... Only what there is of an object in this way, is taken to be existing... It is in Descartes' metaphysics that, for the first time, being was determined as the object of representation ("*Gegenständlichkeit des Vorstellens*") and truth as the certainty of representation ("*Gewissheit des Vorstellens*")... All modern metaphysics, including Nietzsche, goes in the way of this interpretation of being and truth that was initiated by Descartes".[4] According to Heidegger, this metaphysical tenet determines not only the essential features of science, but it also deeply influences the whole life of modern man. Because the objective scientific thought allegedly is always necessarily alienating and reifying, modern man must feel himself homeless in the world, in so far as his consiousness is moulded by objective scientific thought. Therefore Heidegger thinks that the only escape from this human alienation and homelessnes is abandoning the path of scientifically objective thought and passing to what he calls the "different", the "future" thought, and what, in its content and form, is a sort of existantially-mythical thinking, the world of which is the inseparable Four of Earth and Heaven, Gods and Mortal Men.[5]

Hence, Heidegger's attack is directed against all theoretical thought which proceeds in an abstract opposite of subject and object, thought and being,[6] and therefore also in a mutually self-supporting polarity of subjectivism and objectivism. As distinguished from Husserl, who regarded physicalist objectivism as the only source of what he called the "crisis of European science", Heidegger (after "the turning") sees the situation as being more complicated: "The interconnection and mutual impact of subjectivism and objectivism is fundamental here."[7]

The Marxist critique of the Cartesian framework agrees with Heidegger in assuming that thinking remaining in unmediated abstract opposition of subject and object, hence in the polarity of subjectivism and objectivism,[8] cannot meet our needs — in spite of many actual and potential successes of abstract technological thought in several branches of technological application. Yet while Heidegger is proposing a regressive outcome by generally abandoning a scientifically objective approach, the dialectical-materialist critique of the Cartesian framework criticizes and abandons physicalist objectivism for its *not being objective enough* and aims at deepening the *objectivity* of thought.

The mode of thought, the kernel of which is materialist dialectics, is more objective than the previous approaches, mainly because it joins the general conception of a processual character of all being with the discovery of the practical basis of thought and humanity. Both are indispensable to the rise of the new mode of scientifically objective thought. In it, the notion of what there actually is turns out to be richer and deeper not only because objective natural motion is regarded as qualitative variability and development, but also because the processing unity of subject and object becomes the target of *objective* theoretical reflection. Attention to the study of the coincidence of man's changing world and man's self-changing is one of the fundamental features of the new mode of thought.[9]

The new mode of thought overcomes the *objectivistic* tenets by comprehending that men are at the same time the playwrights and actors of their own drama and that all cognitive thinking is part of this practical life process. Simultaneously, the new mode of thought is fundamentally critical of all kinds of *subjectivistic* tenets[10] which, in one way or another, deny the possibility of objectively true thinking, i.e. of grasping by reflective thought — to an increasing extent — the very nature of what is actually going on in nature and society. Subjectivism hinders actual human emancipation by cultivating illusions that idealistically distort man's intellectual and practical self-determination, and, therefore, expose man to unmastered dependencies upon unknown objective factors of natural and social reality.

The new mode of objective thought is open to further development. The intellectual reproduction (and partly pre-production) of what is going on can never be closed or complete, it always develops as the actual reality also does. This implies for the conception of dialectical-materialist determinism that not only the strict mechanistic (Laplacean) form of determinism in its absolutized version is abandoned but, moreover, it makes us ready to accept and acknowledge also the new, not yet known, modes of connection and change, as soon as they are proved by experience, that is not to stick dogmatically only to those forms of determination that have been discovered up to now.[11]

As to the logical-ontological forms, the new mode of thought has been substantially prepared by their historization consisting in a transition from thinking in fixed opposites to thinking that is aware of the fact that the polar antagonisms and opposites, though to be found in nature and society, are only of relative validity, and that their imagined rigidity and absolute validity have been introduced into nature and society by our reflective minds. One of the main preconditions of the new mode of thought is, therefore, making opposites flexible, i.e. the dialectical relativization of opposites. This implies that on the basis of the historified asnwer to the question of what "to be" means, also e.g. the categories "thing" and "essence", along with other categories, are modified so as to express aspects of a process which is finally the process of self-production. Accordingly, the notions of proof and explanation together with some forms of predication and of rational argument also change and are modified.[12]

In a sense, the new mode of thought seems to be closer to Aristotle's conception of ON and OYSIA than to the conception of "the thing" with which modern mathematical natural science started. For Descartes and Newton the natural *ens*, the natural thing, was indentified with *corpus*, subject to the law of intertia. The difference between Descartes' and Newton's views on this question — that e.g. according to Descartes gravity does not belong to the general objective necessary properties of every natural thing — are unimportant here, for they are differences within the same basic historical type of rationality. Most significant here is having natural things rid of self-movement[13], or reducing self-movement to *motus localis* in the same direction and in a straight line. The new mode of thought with its fundamental conception of things as a process and of motion as qualitative change, appears to go back to Aristotelianism. At the same time the dialectical-materialist type of rationality, which is historical and practical, is of course, irretrievably distant from the Aristotelian one, which was contemplative and non-technological. For the new mode of thought comes into being — as far as theoretical forms are concerned — as an outcome of the development of modern science itself; it is inconceivable without the intellectual results of several post-Galilean centuries and is prolonging, in critical continuity, this intellectual work. It keeps approaching experimentally the reality with the aim of transforming it and develops this tradition into new forms. Philosophically its rise is mediated, in the main, by the German classical philosophy, by the philosophical evolution from Kant throught Hegel[14] to Marx.

It seems generally advisable not to exaggerate the novelty of the historical types of rationality and not to ignore the continuity while appropriately stressing the discontinuity. The relationship of the dialectical-materialist type of rationality to the Aristotelian and Cartesian, appears as a historical negation of negation, which means the critical overpassing, but also the critical continuation of the intellectual work done by both.

At this point, too, Heidegger's and the Marxist dialectical-materialist critique of the Cartesian framework (and of the whole tradition of theoretical thought) fundamentally differ. Although Heidegger clarifies explicitly in many respects the difference between ancient, medieval and the modern conception of what "the thing" is,[15] he subordinates these differentiations to his key conception, according to which the basic logical-ontological attitude, as expressed in the Cartesian scheme, has its roots in the Platonic-Aristotelian ontology and, in spite of a new initiative, moves within the limits of the Aristotelian conception of ON and OYSIA. Trying to forcibly accomodate the explanation of these Aristotelian ideas to his own conception of a so-called ontological difference, Heidegger says that by these terms Aristotle expressed only a finished pre--standing given objectivity.[16] This allegedly leads to metaphysical thinking, which moves within an abstract opposite of the subject and object and takes the subject for a fixed thing, considering *logos* as a fixed reflection of a thing. Both poles of this opposite have also allegedly acquired the character of something pre-given (*Vorhandensein*), which is neither a moment nor a product of self-production, all being started to be considered as an object of predication, as "*das jeweils Zugrundeliegende, das Vorliegende.*"[17]

But in fact Aristotle, in his speculations, did not only ask about the mode of being of what is presented to us in the shape of a finished objective givenness (which is "*das Vorhandene*"). In his strictly secular philosophy he mostly understood very clearly the notion OYSIA as an expression of the moving essence[18] of all being. The trends of a dialectically self-developing and at the same time articulately formed conception of being, contained in Aristotle's original thought, are pushed into the background by Heidegger. Thus he constructs a fixed, far-fetched opposite between the thought of pre-Socratic thinkers and Aristotle.

Heidegger's consideration that both Aristotle and Descartes explain the being in essence identically arises most probably out of the fact that he measures them by his far-fetched conception of the so-called ontological difference. And since he obviously does not find with either of them an idea of existence in Kierkegaard's sense, he imagines them as thinkers in the same category of "objective" thinking which has to be overcome.

However, we can partially justify Heidegger's opinion that all being started to be considered as an object of predication, as "*das jeweils Zugrundeliegende, das Vorliegende*" not as late as in the Cartesian framework but as early as in Aristotle's. We can do this if we set apart diverse forms of predication used by Aristotle in his real contemplation on various forms of being e.g. in his *Metaphysics*, and confine ourselves to his explicit theory of predication, as presented in his work *On Interpretation.* In this work he understands the categories *thing-quality-relation* in a way compatible with the Cartesian framework. Therefore this version of Aristotle's explicit theory of predication can really be understood as essentially identical with any of those theories that consider, within the limits of the Cartesian framework, anything as a pre-given pre-standing fixed object which bears certain qualities and becomes involved in various external relations.[19]

This partial justification of Heidegger's diagnosis does not change the fact that — summing up — it is necessary to characterize Heidegger's critique of the Cartesian scheme as a reactionary romantic mystification of the critique of theoretical thinking, moving in the abstract opposite of subject and object i.e. of merely observing and abstract technical thought, accompanied by subjectivist anthropology as a complement to it.

Heidegger's critique starts from the true (negative) recognition that a merely observing consciousness cannot be the last foundation of thought and truth and that cognitive thinking is always a moment of human life, the essence of which is by no means a simple observative experience. But Heidegger reduces at the same time the active essence of man and thinking (i.e. practical being and the practical basis of thinking and humanity) to *existential being*, i.e.

again only to a sort of conscious mental experience and mere "internal" acting. This mystifies the practical basis of thinking and humanity. Heidegger's critique of the Cartesian framework thus becomes in a confusing unity not only a critique of merely observatively objective and abstract technical thinking, but also generally the critique of any *objective* thinking including the one which no longer proceeds in the abstract opposite of subject and object, subjectivism and objectivism, but on a new ground of rationality, conditioned by an understanding of the historical-practical basis of thinking and humanity.

Heidegger's critique *starts* with an important claim to "think more originally than metaphysics",[20] but it *results* in a reactionary romantic answer, which is an extrapolating sophism: since scientific rationality of a certain historical form (i.e. the Cartesian framework, juxtaposing subject and object against one another) is not sufficient for truthful thinking of man to-day, it means that *no* scientific rationality is sufficient and that it is necessary to resort to existential-mythical thinking.[21]

This extrapolating sophism also lies behind Heidegger's reflections on the problem of homelessness, which implements at the same time a certain conception of the relation between rationality and morality.

Since Heidegger deliberately gives absolute validity to certain transient forms of the scientific approach and presents them as the scientific approach in general, he can say that the human being necessarily separates himself from the world when taking an objective theoretical attitude and thus the world ceases to serve man as his home and turns to a subordinate object, a subject of cold observation and analyses, a game of objective forces, which serve man but leave his heart empty. In order to excape from this homelessness and find his home in the world, man, according to Heidegger, has to abandon a scientific approach as a whole, i.e. scientific "objective" thinking.[22]

Heidegger's statement that objective scientific thinking drives man to homelessness, is a mystifying half-truth. Its truthful aspect lies in the fact that if scientific thinking moves within the abstract opposite of subject and object, it cannot provide satisfying orientation to the unification of rational and moral behaviour. The mystifying aspect lies in the fact that it excludes, *a priori*, the existence and even the possibility of such a form of objective scientific thinking, which overcomes the polarity of subjectivism and objectivism and remains at the same time scientific thinking.

In these questions, too, the dialectical-materialist critique of the Cartesian framework has taken a fundamentally different path since the rise of Marxism.

One of the essential features of the dialectical-materialist type of rationality is that it offers — through its very *deeper objectivity* conditioned by the double historization mentioned above[23] — an understanding of the world as man's home and is oriented to the unification of rationality and morality.[24]

If in the Cartesian type of scientific rationality the categorial equipment and the conception of predication and modes of reasoning are so narrowed that logical-ontological categories of a fixed state together with categories of mathematically-conceived mechanical motion are considered to be primary and processes of becoming (together with predications expressing becoming) are considered to be secondary, which can be reduced to fixed states and mechanical motion, and can be derived from them — then, of course, such a type of scientific rationality must leave man in the lurch on moral questions or necessarily invites man to look for support of morality in other spheres than in scientific truth, i.e. usually in religious belief, etc.

By way of illustration we can point to a thinker who contributed in an outstanding way to the clarification of logical foundations of the mathematizing type of science. In § 183 of his "*Wissenschaftslehre*", B. Bolzano noticed a remarkable kind of sentences — as he says — which express becoming *(ein Werden)*. He wanted to clarify the logical character of these sentences by transforming them to a shape presuming a non-self-developmental conception of change. He believed that the sentences "*M* arises from *A*" ("*M* wird aus *A*") or "*A* turns into *M*" ("*A* geht über in *M*") could be changed into an allegedly more exact logical form: "The subject *A* has a property such that it undergoes a change, the effect of which is that in future it will become the subject *M*" ("Der Gegenstand *A* hat die Beschaffenheit, einer Veränderung zu unterliegen, deren Wirkung ist, dass er in einer künftigen Zeit *M* sein wird.").

The incapability of unifying scientific rationality and morality on the platform of secular philosophy is shown in an interesting way by Bolzano's work "*Wissenschaftlehre und Religionswissenschaft in einer beurteilenden Übersicht*" (Sulzbach 1841). The strict logician thus bases morality in the end upon the authority of religious revelation. As to the type, this is the same as can be found in the famous *Scholium generale* of Newton's *Principia Mathematica Philosophiae Naturalis*, in which a massive theist theology is regarded as a logically necessary complement of the mathematical-mechanical image of "*constitutio mundi*".

The dialectical-materialist type of rationality can offer — precisely through its deeper objectivity conditioned by double historization — a comprehension of the world as man's home and can be oriented to unification of rational and moral behaviour, since it is a theoretical expression[25] and organ of a real historical movement, of a revolutionary change of capitalist society into a communist one, and therefore an organ of a new historical form of humanity being born in this revolutionary struggle for the reorganization of society.[26] For this historical form of humanity and for this new thinking as well, the world becomes man's home because it ceases to be an object of exploitation and control from the viewpoint of such utilitarian estimation, which is essentially determined by commodity relations. The criterion of utility and of control (including self-control) acquires a different historical quality since they become a subordinate moment of a real humanistic life attitude. Real humanism takes as ist highest moral value and self-purpose the fullest possible development and the self-realization of all people as socialist personalities in cooperation, friendly co-existence, with the help of self-criticism. This self-purpose is always immersed in real human history and modified by historical movement.[27]

As to the appropriation of the world, the difference between new objective thinking and the Cartesian thinking does not lie n the fact that the latter is oriented to gaining mastery over natural forces, while the former is not. The difference lies in the way in which the orientation to the mastery of natural forces, common — in a different historical quality — to both Cartesian and new thinking, is interwoven with the whole human life process. There is a difference between utility and mastering, determined finally by the rule of commodity relations, and the utility and mastering, determined by conditions of a real humanistic life attitude. For this reason the new thinking uses a modified conception of technology in comparison with Cartesian-type of hinking, because technical thinking becomes a subordinate part of historical-practical thinking. Only in relatively narrow limits can technology be inidependent of historical-practical social circumstances and can be understood as mere application of a pure theoretical science. In a larger historical-social context, technology and technical thinking become a moment of the totality of social life. A certain form of technology is connected with a certain form of social life.

Moreover, the new objective thinking is but one of the aspects of man's real humanistic appropriation of the world — a form of theoretical-intellectual appropriation. Other forms, some of them basic, some derived, are the material, practical appropriation (productive, including technical), aesthetic, moral, emotional appropriation, etc.

A special role is played by the complementarity of scientific (including scientific-philosophical) thinking and literary artistic thought, with their permanent difference and mutual influencing. Aristotle's opinion that poetic thinking is more truthful than theoretical,[28] can be accepted in the historical modification that mutual complementation and stimulation of scientific and literary artistic thinking has growing importance for real humanistic self--development of man by man by means of many-sided appropriation and a transformation of the world. Forms of literary artistic appropriation of the world (like all forms of appropriating the world by artistic creation) gain importance without reducing the great and indispensable role of objective theoretical thinking. One can foresee that in the communist way of life the complementarity of new objective thought and literary artistic thinking will be reflected more and more effectively in the communicative thinking of everyday dialogues.

If our sketch is not mistaken it would imply that the mode of thought, the kernel of which is materialist dialectics, links creativity and science in many aspects and dimensions. Even the Cartesian framework was an important turning point in the path of human emancipation from the Christian religious conception of creativity. Nevertheless, in its explanation of being it remained, in many respects, on the level of this conception. We have tried to argue that real humanism profoundly secularizes creativity because — in connection with new historical circumstances and human problems — it is essentially connected with what we see emerging in our epoch: with the new historical type of scientific rationality.

REFERENCES

1 In our view, reason originates and develops as a social human characteristic, in inseparable context both with work and speech, and has a biosomathical carrier; it is bound up with the brain and nervous system of living man in social communities. Reason is that habitual property of human beings, always socially united in concrete, historically developing forms, which shows itself a) in the ability to create, to justifiably unite and disunite meanings and to evaluate them as to their truth; b) in the ability to carry out purposeful actions aimed at satisfying the material and spiritual needs by realizing preconceived goals by useful means. Rational thought is movement of meanings in a five-dimensional field.

2 Pre-eminent = in comparison with others remarkably influential on a large scale and for a long period.

3 In his "Philosophical Notebook" (*Works*, vol. 29, p. 202, Russian ed.), V. I. Lenin mentions "the objectivity of treatment" in the first place among the elements of dialectics.

4 M. Heidegger: *Holzwege*, 4 th ed., Frankfurt 1963, p. 80.

5 Cf. e.g. M. Heidegger: Das Ding. In: *Vorträge und Aufsätze*, Pfullingen 1954.

6 In Heidegger's "*Einführung in die Metaphysik*" (3 rd ed., Tübingen 1966, pp. 88–89), the very nature of metaphysics is characterized as „die massgebende Herrschaft der Scheidung 'Sein und Denken' im abendländischen Dasein". „Das Denken setzt sich dem Sein dergestalt gegenüber, dass dieses ihm vor-gestellt wird und demzufolge wie ein Gegen-stand entgegensteht... ...Wir müssen in der scheinbar gleichgültigen Scheidung *Sein und Denken* jene Grundstellung des Geistes des Abendlandes erkennen, der unser eigentlicher Angriff gilt.".

7 M. Heidegger: *Holzwege*, l.c., p. 81.

8 That is thinking in a spectator's position and abstract technological thinking, accompanied by various forms of subjectivist anthropology as a complementary opposite.

9 For the sake of brevity her and below the term "new thought" or "the new mode of thought" will always be an abbreviation meaning themode of thought, the kernel of which is materialist dialectics, i.e. thinking within the framework of the dialectical-materialist type of rationality.

10 Cf. P. Fedoseyev: "Leninism versus Subjectivism and Objectivism", in: *Social Sciences Today*, Editorial Board USSR Academy of Sciences, No. 5/30/, 1974.

11 Cf. V. I. Lenin's comments on philosophical views of some excellent natural scientists

that they "conceive the idea of change in a rather restrictive way". (*Works*, vol. 38, p. 269, Czech edition).

12 For more details, see J. Zelený: *Wissenschaftslogik bei Marx und "Das Kapital"*, Akademie-Verlag, Berlin 1968. The English edition has been published by Basil Blackwell Oxford 1980 under the title *The Logic of Marx*. Some problems of rationality types are discussed in *Studien über Dialektik*, Acta Universitas Carolinae Monographia LVII, Praha 1975.

13 We refer here to Descartes' and Newton's principal conceptions of motion and the natural thing that entered into the foundations of the rising modern mathematical natural science. In the original works of Descartes and Newton — as usually is the case with great thinkers—one can find anticipatory seeds of other conceptions transcending what was accepted and became widespread in the given period. For the concept of motion and change, cf. e.g. § 45 in Descartes' *Principia philosopphiae* (3 rd part) and Query 30 in Newton's *Optics*.

14 Hegel's view that everything depends on grasping and expressing the truth not as substance only, but as subject as well, and that "the living substance... is truly realized and actual solely in the process of positing itself, or in mediating with its own self transitions from one state or position to the opposite",—i.e. Hegel's conception of being as a process of self-movement and self-becoming—was a milestone in the historical transition from the Cartesian to the dialectical-materialist type of rationality. (Cf. *The Phenomenology of Mind*, transl. by J. B. Baillie, London 1910, pp. 16—16.)

15 E.g. in his work *Die Frage nach dem Ding. Zu KantsLehre von den transzendentalen Grundsätzen*, Tübingen 1962, pp. 62 and ff.

16 I.e. what Descartes calls "obiecta proposita" in *Principia philosophiae*. Compare with M. Heidegger: *Einführung in die Metaphysik*, Tübingen 1966 (3 rd ed.), p. 147: "Es (i.e. the word OYSIA—J. Z.) meint das Sein im Sinne der ständigen Anwesenheit, Vorhandenheit", ibid., p. 154: "Bleiben, Immergleichheit, Vorhandenheit, Vorliegen—sagen im Grunde alle dasselbe: ständige Anwesenheit: ON als OYSIA."

17 Ibid., p. 142.

18 Aristotle's OYSIA is more "the non-incidental way of being of what is concerned" than the Middle Ages *essentia*.

19 Even Frege's critique of Aristotle's theory of predication, as given e.g. in "Nachgelassene Schriften" (ed. Hermes-Kambartel-Kaulbach, Hamburg 1969, p. 16), relates in fact to a narrowed, simplified Aristotle. From the point of view of the hypothesis of historical types of scientific rationaliy the place of Frege's theory of predication appears as follows: In spite of epochal innovations, connected with the introduction of a generalized conception of function into logic, with the conception of predication as filling up of unfilled functions, etc., Frege's theory of predication and inference moves in the field of the Carthesian type of scientific rationality. Frege strictly rejects the historization of logical problems (ibid., p. 159 etc.). This does not reduce Frege's merits in clarifying logical structure and the nature of some aspects of mathematical knowledge. Since the mathematically conceivable side of all things and processes is an important aspect of the new objective thinking as well, Frege's discoveries remain relevant even for the new historical type of rationality While Heidegger considers mathematical logic to be "no logic" (*Die Frage nach dem Ding*, Tübingen 1962, p. 122), for the dialecticalmaterialist type of rationality it is an important, although subordinate, *fragment* of logic.

20 M. Heidegger: *Über den Humanismus*, l.c., p. 36.

[21] An analogical kernel can be found in other contemporary conceptions, which ,in spite of different argumentation and erudition, recognize the crisis of bourgeois culture and —remaining within the horizon of bourgeois society—look for a possible outcome. E.g. Daniel Bell, acknowledging that "we are witnessing the end of the bourgeois idea" (p. 7), asks in his work "The Cultural Contradictions of Capitalism" (London 1976, p. 29): "What holds one to reality, if one's secular system of meanings proves to be an illusion?" He answers: "I will risk an unfashionable answer—the return in Western society of some conception of religion". Bell cannot imagine the rational secular system of meanings in other form than a "technological" or "functional" rationality within the limits of the Cartesian framework.

[22] Claiming to criticize and surpass all alienating thoght, which "does not sufficiently highly" estimate and emphasize man's dignity (*Über den Humanismus*, l.c. p. 19), Heidegger offers, in fact, in a fictitious "elevation" of man above all the material, a philosophy of human weakness, passivity, human degradation. Existential mythical thinking has an ideological function as to objective reality, it supports the idea of "leaving thinfs as they are", "non-interference" ("Gelassenheit").

[23] I.e. historization in the sense of the processual conception of all being and historization in the sense of discovery of the historical-practical basis of thought and humanity.

[24] The acts of human life have a moral character if they are motivated by a conscious effort to achieve what is taken to be good and to avoid and overcome evil. This effort is accompanied by moral value judgements and feelings of consciousness. And what is of greatest interest for us here: this effort can rely to various degrees upon rational reasoning of human acts, or more generally: it can, in various ways, be connected with a rational reasoning of human acting and with a rational explanation of reality.

[25] This is not to be understood in the sense of sociological relativism as professed e.g. by A. Sohn-Rethel. A criticism of sociological relativism, which exaggerates the linking of theoretical forms to the forms of social practice, is given in our article "Vom sozialen Wesen der Erkenntnis", in: *Aspekte der materialistischen Dialektik*, Edition UFS, Prague 1977.

[26] This and other important aspects of historization of the forms of being and thinking are treated in Radovan Richta's essay "Is There a New Type of Science Emerging", in: *Theory of the Development of Science*, No. 1, Prague 1977.

[27] The conception of a real humanistic morality does not make sense without a far-reaching rational explanation of a given reality in its law-governed development. It presupposes Marx's conception of scientific analysis of social reality, claiming that: As soon as a scientific theory expresses what is going on it ceases to be doctrinaire and becomes revolutionary. (*Werke*, vol. 4, p. 143). It stops being doctrinaire, i.e. merely descriptive, indifferent to conflicts of values, and becomes revolutionary, i.e. orients necessarily to a certain practical, including a moral-practical, evaluation. Not every isolated—although truthful—predication of partial facts and partial processes is value-impregnated. As an isolated predication it can, in the moral respect, be completely free of any evaluation. In order to become the pole of a new unification of rationality and morality, scientific truth has to be a theoretical cognition that a) does not stop at isolated partial predications but strives to unite and integrate them into apossible complex theoretical reproduction of interconnections of nature and society, taking into account the dialectics of the absolute and the relative; b) is conscious of its reflective character with respect to objective reality but does not remain at the conteplative opposite of "thinking" versus "objective reality"

and proceeds to assuming scientific truth to be a moment of a human (social) way of life.—Why do we speak about a "new" unification? Since it is not the first case in history when rationality and morality have been united. Even the Aristotelian type of rationality unified in its own way rationality and morality, it was "agathological". In the modern age, Spinoza unified rationality and morality in a specific way as well—and, of course, so did the great dialectician Hegel.

[28] Aristotle: *Poetics*, 1451 b.

6. Integration of Cognition and Materialist Monism

JIŘÍ ZEMAN

Contemporary philosophical materialism proceeds from the assumption of the material unity of the world. Especially today, in the age of unprecedented information growth, at a time when scientific cognition (including individual areas of knowledge) is becoming a matter of ever increasing differentiation and complexity, that the principle of materialist monism acquires special methodological relevance. The principle of monism applies to the objective character of material reality as well as its reflection in cognition. It is from this particular methodological viewpoint that scientific philosophy approaches problems affecting the integration of scientific cognition. This wide-ranging problem area has a relevant bearing on the principal issues confronting the theory of materialist dialectics, the theory of reflection, the systems approach, the clarification of the specific nature of different forms of motion, the structural-genetic unity of the world, etc.

Integration of cognition is connected with problems concerning the relation of philosophy and specialized sciences, the unification of scientific disciplines, their laws, terminology, etc. Philosophy and specialized sciences were originally coextensive with philosophers acting as scholars and specialized scientists at the same time. Gradually, however, individual branches of science emerged gaining autonomous status. Nevertheless, this differentiation did not imply a gradual elimination of philosophy. Apart from many other functions, philosophy has retained its legitimate integration function. Without interfering with the differentiated nature of cognition, scientific philosophy has a synthetic character, an integrating and unifying force. While making growing inroads into all areas of science, materialism and dialectics point to the profound interdependence of all concepts and all spheres of cognition. The principle envisaging the world's material unity postulates a common basis for all phenomena and the ultimate interdependence of all differences and varieties. The dialectical character of reality thus presupposes a dialectical unity of counterposed phenomena. This fact is duly reflected in the development of modern science, which mirrors the process of linkage of different concepts, laws, regularities, etc. The differentiation process is accompanied by integration processes traceable even in 19th and 20th century science, which witnesses the linkage of different fields of study and the rise of interdisciplinary sciences. Points of contact emerged between mathematics and logic giving rise to mathematical logic, between mathematics and geometry in the form of analytical functions, between geometry and mechanics in the theory of relativity, between physiology and psychology in neurophysiology, between physics and chemistry in physical chemistry, between chemistry and biology in biochemistry, between physics and biology in biophysics and bionics, between logic and linguistics in semiotics, etc. Integration of cognition is also a feature characterizing the combination of several branches simultaneously, e.g. cybernetics, systems theory, etc. The unification process is, of course, far from easy, as is evident from the difficulties attendant upon the linkage of quantum theory with the theory of relativity, etc.

The development of modern science reveals the gradual accumulation of knowledge concerning the dialectical character of phenomena and the dialectical interdependence of counterposed concepts. In mathematics, this applies, e.g. to the study of the continuity/discontinuity problem, in physics — the specification of the dialectical interdependence of space and time, mass and energy, wave and particle, in biology — the dialectical unity of organism and environment, in neurophysiology — the dialectics of excitation and inhibition, structure and function, etc. The integration process in science is determined by the material unity of the world and the dialectical character of phenomena. Drawing a sharp line of demarcation between counterposed phenomena, (the essential and the accidental, space and time, etc.) metaphysical conceptions of thought viewed a specific opposition as a kind of zero function of another opposition, briefly, as independent of this opposition. It is becoming clear, however, that opposites pervade and permeate each other, i.e. display a certain functional interdependence, determine each other and the mutual boundaries are by no means sharp. What is needed is the specification of the type of functional dependence in question. The endeavour to explore both their content as well as their formal (e.g. mathematical) aspect cannot be written off as irrelevant, despite the fact that science has not yet surpassed the initial stage of inquiry into the dialectical unity of the content and form of these concepts.

The envisaged integration of scientific cognition must study not only the content but also the form of scientific findings, i.e. the language of science — scientific expression in a sense duly including the formal-analytical descriptions of phenomena. The two aspects under consideration were given their first theoretical basis in Leibniz's formulation (universal characteristics, combinatorics, encyclopedia). Today, with the increasing evidence testifying to an ever more profound content unity of science and, what is more, the emerging possibilities for the attainment of formal unity, see the unified mode of communication (information languages, semiotics, mathematical logic, mathematics), this endeavour acquires a more definite form. Although ventures in this particular field are still in their infancy, the claim that dialectics is diametrically different from mathematics and logic is untenable. The linkage of the two is unassailable on both grounds. The unification of sciences and scientific language cannot, however, be effected solely with regard to form, or — more precisely — in terms of the narrow interpretation propounded by neo-positivism (*Einheitswissenschaft*, *Universalsprache*). This kind of reductionism took recognizable shape in so-called physicalism, especially in the works of Carnap and Neurath in the 1930s. Physicalism attempted to translate all concepts specific to different sciences in terms of the notional apparatus current in physicalist speech. The language of physicalism purported to employ terms and concepts referring to things ("red", "table", "next to"), to make this communication more precise and introduce new terms (e.g. "mass", "energy", etc.). The remaining sciences, it was argued, proceed from this language and contribute further concepts (such as "organism", "thinking", "state", etc.). The conclusion offered was that all concepts current in other sciences can be translated into physicalist speech (i.e. concepts specific to sociological sciences are first translated into psychological concepts, psychological concepts into biological concepts, and finally these are translated into physicalist speech). This led to the reduction of superimposed levels to subordinate levels and the concurrent obliteration of qualitative specificities. Although physicalism was subsequently abandoned, it still characterizes positivist thought.

It is clear that the unification of scientific cognition cannot be accomplished solely on the basis of formal and quantitative criteria, without regard to content and quality aspects. Science is not characterized by a merely formal arrangement of a system of propositions, but ultimately reflects objective reality. The criterional content of science is objective, unified, material reality; it is in this reality that the foundations of the unity of knowledge and the possibility of its integration and unification reside.

Scientific materialism is a prerequisite for the formulation of a general, universal system of representation, and/or the provision of an information system of a higher magnitude. Different phenomenal areas can be represented in a different manner and with a varying degree of adequacy. Priority should be assigned to a system of representation that is more accurate and at the same time more general. This principle proved its validity during the confrontation of the geocentric and the heliocentric system, Euclidean and non-Euclidean geometry, classical and non-classical physics, etc. Apart from functioning as a universal, unifying theory and method (as a mode of explanation of phenomena and a guide towards their cognition and mastery), dialectics likewise offers a viable basis for the integration of different fields of cognition as a universal representation system, thanks to its reliance on the fundamental principles governing the actual reality represented — these principles include infiniteness, unity, diversity and mobility. Dialectics constitutes an open-ended system allowing for a relative finalization of individual stages of cognition and for the further, infinite advance of the process of cognition towards the assumed absolute truth.

In the diversity of facts and findings it is thus possible to seek unity and interdependence, to strive for an ever more perfect order, for the elimination of imprecisions and errors. An information, representation system must respect the content aspect of the phenomena under description. At the same time, this system must be endowed with an adequate mode of expression, i.e. description of the content; it must duly respect the formal aspect, i.e. the "language of science" which is studied by semiotics, science methodology, logic, etc. In its further development, dialectics must therefore apply this criterion in the study of formal-analytical, logical, mathematical and semiotic means, if it is to go beyond the confines of a discipline that is of purely general-methodological and qualitative-content character. The requirements of dialectics include, *inter alia,* such factors as unity of form and content, quantity and quality, etc. This objective must be pursued both by philosophy and the specialized sciences within this specific range of appurtenance.

Important stimuli for a correct understanding of the problem of integration of scientific cognition and the character of contemporary science, especially natural science, are likewise provided by the principle of simplicity which together with other factors such as invariance, symmetry, etc. plays an important methodological role in this particular problem area. The simplicity principle helps to shed light on the character of rationality, i.e. the mode of thinking specific to contemporary natural science, as well as the issue of determinism.

The problem of the world's unity (reflected in a complex manner in the unity of cognition) is connected with the problem concerning the character of laws and regularities governing objective phenomena and their reflection in cognition. Cognition in the field of natural sciences strives for exactitude, for the representation of objective determinism, for unequivocality. This type of cognition excludes subjective arbitrary biases. True, the subject selects specific aspects of reality essential for a successful realization of human practice according to the character of human qualities and social needs, but even this choice cannot disregard the determinist character of objective natural (or social) laws. Objective laws are unambiguous. Cognition in natural science is based on the requirement of unequivocality, which is to a considerable extent related to the simplicity requirement. Specific phenomena or processes may display considerable diversity and complexity, yet their realization must always comply with principles admitting of only one real solution: the becoming actual of what was potential, i.e. transition from the field of the possible to the field of reality, which is concrete, actual and unique — this occurs on a law-governed basis, i.e. inevitably and irreversibly (the laws of probability, too, are objective and deterministic). The idea of simplicity, unequivocality and determinism in natural processes was heralded way back in the history of new age science by scholars such as Newton, Leibniz, Lomonosov, and others.

The simplicity principle derives from physics. In the 17th century, Fermat formulated the principle of minimum time and minimum trajectory (light and bodies travel the shortest track within the shortest possible time interval). In the 18th century Maupertuis formulated the principle of minimum action (minimum operation) as follows: bodies move so as to reduce the general amount of strain involved in the action to a minimum, i.e. nature uses the simplest means possible (the effort required for the transference of a body is, according to Maupertuis, equivalent to the product of minimum trajectory, minimum time and minimum exertion of energy). These ideas were further developed in the so-called variance principles of mechanics, which in fact formulate the most general laws of physics. Of all the possible events of a specific class only those are realized in which a certain quantity acquires extreme value. Elaborated by d'Alembert and Euler, these principles were subsequently detailed by Lagrange and Hamilton and finalized by Poincaré, Lie, Hilbert, Noether, and others.

The simplicity principle was also applied in quantum mechanics. Defining the quantum of action in his constant h, Planck regarded the quantum action principle as the cornerstone of a unified physical picture of the world. The same line of endeavour was pursued in studies of indefiniteness, in the unified field theory, the principle of complementarity and in other tentative formulations of a general theory of physics. Further, the idea of geodetic curves (the shortest connectives) and spatial curvature initiated an analogous line of speculation in relativity theory which saw the development in terms of the quantum action principle of the idea of the shortest interval, i.e. the minimal (invariant) distance between two points-events in space-time.

In Minkowski's curved four-dimensional continuum it is the shortest interval that provide the basis of all events. The distance of a point (event) from a specific inception (the chosen spatial-temporal present) is for this continuum formulated as

$$s^2 = c^2t^2 - x^2 - y^2 - z^2.$$

This spatial-temporal interval (and/or quadrate of the interval) is invariant here. $s^2 = 0$ would involve a line of zero length (light signal of velocity c). In Minkowski's 'light cone', s in the future and the past is greater than zero (i.e. the interval is positive). If the inception is characterized as $t = 0$, then in the field of the future t is greater than zero and in the field of the past t is smaller than zero. The light cone is delimited by lines with $s^2 = 0$. Apart from these fields there is the extraneous area (the so-called quasi-present, in which velocity c would be exceeded), where s^2 is smaller than zero, i.e. the interval is negative. In the real world, spatial-temporal intervals are positive, i.e. s^2 is greater than zero. Hence the positive intervals in time (i.e. positive time flow, positive interval between two events, which cease to be identical), positive length, positive dilution of mass, positive entropy increase, etc.

Methodologically, cybernetics, too, proceeds from the simplicity principle, e.g. in Ashby's formulation of the essential variety of a functioning system, etc. Simplicity principles also relate to other important ideas of modern physics and science in general, such as the questions of invariantness, symmetry, etc. The simplicity principle has its objective and subjective, ontological and epistemological aspects and asserts its force both in nature and in thinking. An incorrect interpretation of the simplicity principle characterizes teleologism, which regarded simplicity as an expression of supranatural intelligence (e.g. Leibniz's pre-established harmony) — instead of viewing it as a natural expression of determinism. At present, a positivist-subjectivist interpretation seems to prevail. This interpretation is founded on the tenets postulated by Avenarius and Mach accentuating the so-called economy of thought. According to Mach, physical laws of conservation are a corollary of the economy of thought which has actually no objective basis but is of biological-psychological character (goals are attained by exerting minimum intellectual effort). This subjectivization of science is further developed in the works of some Western scientists. According to Goodman: "Rather than the simplicity of science being limited by the simplicity of the world, the simplicity of the world is limited by the simplicity of science."

Subjectivism divorces the objective basis of simplicity from its subjective reflection, which is onesidedly given an absolute status. In a similar fashion, neo-positivism separated the objective unity of the world from the unity of science in the belief that the unification of science and its language can be accomplished on a purely conventional and formal basis. Objective and subjective aspects of unity (just as the objective and subjective aspects of simplicity) do however form a dialectic unity correctly accentuated and accounted for by materialist dialectics.

The simplicity principle is likewise applied in cybernetics. Ashby's principle of the essential variety postulates, roughly speaking, the following: a system capable of homeostasis and development must possess a certain minimal variety. With an excessive variety the system would be unstable, with an inadequate variety the system would be incapable of auto-organization. Hence the principle of essential redundancy can be formulated as follows: if a system is to function (i.e. prove efficient) it must not reach the ideal optimum (i.e. a state without a degradation of information or energy), since this would *de facto* imply a state of equilibrium. The system must possess a certain amount of inefficiency (energy wastage, information redundancy), this being a condition of its func-tioning. According to Prigogine, the imbalance can be a source of order and irreversible processes can induce the rise of a new type of dynamic states of matter and what is described as dissipative structures; the non-linear thermodynamics of irreversible events, he contends, permits a description of the rise of structures after a certain threshold has been transgressed.

According to the redundancy formula, redundancy

$R = \frac{I_{max} - I_{ef}}{I_{max}}$, where $I_{max} - I_{ef} = I_z$; hence utility $U = 1 - R$, i.e.

$$U = \frac{I_{ef}}{I_{max}}$$

I_{max} is supplied information, I_z lost information, I_{ef} effective information. A formal counterpart of U is Carnot's coefficient of thermal machine efficiency:

$$\eta = \frac{Q_1 - Q_2}{Q_1} \text{ i.e. } \frac{A}{Q_1}\text{, where}$$

Q_1 is supplied energy, Q_2 lost energy, i.e. not converted into work, non-effective, A is work, i.e. effective energy.

The principle of essential redundancy proceeding from the principle of quantum action (and/or simplicity) generally applies not only to information processes but also to energy processes (formal analogue of the relation between effective and supplied information and the relation between effective and supplied energy). This principle can be formulated as follows: No real system can possibly function without minimal non-efficiency; in other words, loss-free systems are non-existent, i.e. systems do not function without a certain degradation of information or degradation of energy. The bifurcation into an effective and non-effective component must be minimal — this defines the elementary difference, and/or imbalance.

At the same time, the principle of minimum efficiency is valid: No realy, system can function without minimal (information, energy) efficiency. Every operative material system involves a certain negentropy process (this is also connected with the capacity of auto-organization and reflection).

Modern science frequently substitutes certain minimum or maximum, i.e. minor or major but always final values — e.g. constants h, c, etc. for zero and infinity. This applies equally to the principle of minimum non-efficiency and efficiency, which proceeds from the principle of quantum action, the variance principles and the simplicity principle.

The principle of essential redundancy accounts for the fact that our world is characterized by positive parameters: spatial-temporal intervals, causal processes, thermodynamic clines. All processes represent what is in fact a declination from the envisaged state of equilibrium. Each process is defined by the difference, and/or contrast between two levels and is a cline process (information processes, energy processes, biological processes, psychic processes, etc.).

The principle of simplicity constitutes along with the principle of essential redundancy a characteristic of reality and cognition, it helps to shed light on the nature of contradictoriness in processes, their character and dynamic orientation.

The dialectical materialist conception of this principle is opposed to the religious-teleological and subjective idealist conception. Its further philosophical study, based on the results of modern natural science, is a promising area of inquiry — an area of general methodological validity, characterizing one of the important aspects of contemporary thought in the field of natural science.

Another major contribution to the integration process in cognition was the systems approach, whose emergence in the limelight of modern science was one of the logical spin-offs of the broader process of methodological overhaul affecting contemporary science. The systems approach is linked up with the increasing application of methods and procedures connected with analysis, quantification, mathematization and formalization. The earliest intimations of these approaches, traceable way back to the science of antiquity, gathered momentum with the advent of new age science. It was the 19th century and the advance of exact sciences, especially mathematics, logic and physics and subsequently scientific philosophy, that witnessed the concurrent development of the systems approach. However, it was not until the period following World War II that these new methodological approaches were put on a more systematic basis. Post-war studies in this field focused on the concepts of system and structure and the methods specific to the systems approach, its mode of speculation, etc. Apart from mathematical and logical methods, the so-called general systems theory was gaining ground and the same applies to cybernetics and the rapidly developing structural, relational and structural-genetic approaches. Science began to abandon the traditionally rigorous division into substances and their properties, which strictly separated aspects of substance from aspects of motion. More sophisticated approaches emerged revealing the profound dialectical interdependence of these and similar conceptual opposites. It became clear that the substance substatum cannot be separated from relations, properties and attributes just as the totality of the object cannot be reduced to its manifestations, relations or properties, i.e. to interpretations presenting this totality as a kind of black box characterized by no more than its overt manifestations constituting what is alleged to be the only objective reality. Systems and relational thinking do not constitute the counterpole of dialectical thought. It can function properly and profitably only in conjunction with dialectics. Without this essential complementation, the so-called "pure" systems approach is bound to resort to positivist, relationist, functionalist, structuralist, operationalist and behaviouralist interpretations.

The subject, function and method of scientific philosophy outstep the restricted framework characterizing such approaches — whether they be of the abstract-system, formalist, positivist, structuralist, "purely" analytical or other type. One of the distinctive qualities of dialectics involves the fact that its concern is not restricted to a merely external description of the given phenomena, or — for that matter — to the merely formal aspects of the object under inquiry, or its translation in terms of a symbolic descriptive apparatus. Dialectics is concerned with the objective foundations of phenomena and studies their actual content reaching far beyond the boundaries of subjective reflection. At the same time, dialectics grasps the dialectical unity of the objective and the subjective, the independently existing object and its scientific description, which is after all aimed at practical control and transformation of the object. Dialectics does not refute and indeed cannot refute formal, mathematical and abstract systems approaches. Dialectics comprises, apart from content and quality, also form and quantity. Hegel's dialectics did not rely on pre-eminently mathematical and exact approaches. 19th and 20th century positivists, however, failed to grasp its far-reaching significance. One of the factors stimulating the development of contemporary dialectics is the more consistent and more conscious assimilation of mathematical, cybernetic and systems approaches and methods. The concept of system and structure attains a new qualitative level precisely because the dialectical interpretation dovetails quantitative and non-quantitative, formal and non-formal aspects, etc. Combined with dialectics, the systems approach is therefore of considerable importance for integration procedures in contemporary scientific cognition.

The process of development of dialectical logic, utilizing a number of findings formulated by specialized sciences, their concepts and methods, follows the line of elaboration of Lenin's tenet on the unity of dialectics, logic and epistemology. Lenin placed due emphasis on the need to study the Hegelian problem of the relation between the specific and non-specific, the mediated and the immediate, the historical and the logical, etc. Lenin's conception of logic as knowledge within the entire scope of its development and the specification of the mutually determining inter-relations of all phenomena is of paramount importance. At the same time, Lenin pointed out that the concepts of our subjective dialectics involve a certain kind of simplifying transformation of the objective content, i.e. a translation of the concrete and all-inclusive into the abstract and onesided. Approaches neglecting the objective aspect and confining attention to a merely analytical description of phenomena, relations, etc. stem from the failure to grasp the dialectics of the subjective and the objective.

The problems under consideration are also connected with the issue concerning the relation between substance and form, arrangement and disarrangement, etc. Arrangement, i.e. order, organization, etc. — notions closely connected with systemicness and structuredness relate not only to mere form or description but also to a definite content, to a definite ontic aspect. It is difficult to define an organization or a system. On this issue scholarly opinion is divided with definitions varying from one uathor to another. Essentially, three approaches are possible: 1) the formal-mathematical approach, 2) the natural-scientific approach, 3) the proper philosophical approach. The first type of approach studies structures and systems in an entirely abstract form, the second approach includes aspects of substance and energy, and finally, the third approach studies all these aspects in a general form.

What is important is to understand how the order, i.e. the architecture of the world, is reflected in the order, i.e. the architecture of cognition, the structure of the psychic, the structure of different types of conceptual apparatus, scientific language, and science in general, including philosophy and, further, how the objective is transformed — via the subjective — into practical activity. This problem area is tied up with the relationship of philosophical, general scientific and specialized science concepts. Provided that philosophy captures not only the general but also the particular, what then does the systems approach capture? What is crucial here is obviously the type of interpretation given to this approach, i.e. whether we view this approach in formal psychological terms or in dialectical-systemic terms. In the latter case, one can claim that dialectics develops in that it gradually assimilates exact systemic and mathematical criteria. Dialectics relies on a structural-genetic approach, while also including the temporal-developmental aspect, the objective-content aspect and the quantity-quality aspect. Dialectics duly considers the noetic aspect (of the abstract semantic structures) as well as the ontic aspect (of the objective processual structures).

The current categories of dialectics (interrelation, totality, mutual interaction) are also related to concepts such as organization, system, etc., the concepts are not mutually exclusive. On the contrary, the system of materialist dialectics allows for an organic incorporation of these new concepts. This fact is duly stressed in a number of Soviet studies, e.g. L. A. Petrushenko.

Arrangement, structuredness, information, etc. are concepts that help to develop and advance contemporary materialist monism and, hence, the integration aspects of knowledge. Matter can be viewed as a certain structural and developmental whole. In this interpretation it is imperative to strive for a more precise specification of its general characteristics and individual manifestations. Dialectical development, which is connected with the general ability of self-motion, auto-organization and reflection can be viewed as the *growth of organization.* The classification of forms of motion from low-level types to higher levels is based on the idea of comparing low-level and high-level organized material systems. Today, extensive evidence is available to science pointing to a more detailed characterization of the exponents of such forms of motion as structural substrata of the given functional manifestations.

Organization is not something purely relative, but has an objective character. In rough outline one can study and compare systems and their particular qualitative level and specify the criterion of developmentally lower and developmentally higher levels in the section of universe known to us. The idea of the world's material unity proceeds from the recognition of the objective character of organization. The structure-substance substratum cannot be treated in isolation from its functional manifestations (e.g. manifestations of the energy or information type). Just as essence and quality, structure and function, too, display a dialectical interrelationship and cannot be divorced from each other.

A comparative table can theoretically be constructed in which a certain form of motion is assigned to a certain concrete system as the exponent of motion. This system displays a certain reflection capacity, certain structural, energy and information characteristics, a certain life span, etc. Such systems represent respective developmental quality levels with an increasing degree of organization.[1]

The theory of complex systems accepted by a number of leading Soviet scientists (e.g. the team of authors in S. A. Kaplan's "Extra-terrestrial Civilizations") proceeds from the endeavour to overcome anthropomorphism in human cognition. Man is of course restricted by the conditions of his information sphere including his social anchorage and determination which necessarily constitute his ultimate point of departure. Nevertheless, outstepping the boundaries of the proprieties of space and time, man penetrates further into the micro-cosm and mega-cosm. It is however possible to formulate the *general functional determination of the system* (encompassing life, consciousness, intelligence, society) as a specific methodological model (with due restrictions). In this way, one can ascertain the specific degree of structural complexity, energy output and the information capacity of systems from the lowest types in inorganic nature (connected with the respective forms of motion) to living organisms, man, society and hypothetical super-civilizations. The system performs certain information and target functions, and — at higher levels — homeostatic and control functions. The problem at issue involves optimal system functions, i.e. information functions, control functions, homeostatic functions, etc. Some system parameters can be quantified. The so-called functional determination of a system does not imply that the system is fully determined and defined exclusively by its functions, manifestations, etc. On the contrary, it presupposes that these functions have their exponent. This is not a functionalist or operationalist conception. The functional determination of a system strives for the specification of general characteristics, principles of organization, parameters of functioning and the development of complex systems that could at least be partially quantified.[2] According to dialectics, matter constitutes a scale of systems with an increasing degree of organization. The theory of development of matter towards higher levels cannot regard the issue of organization and its degree as a matter of anthropomorphism. The concepts of system, negentropy, information, etc. are closely connected with the dialectical conception of development as the increasing level of organization in matter. Materialist monism proceeds from a unified classification of forms of motion specific to matter while relying on cybernetic and systemic aspects. The problem of materialist monism and integration of science is thus studied with due reference to the unity of forms of motion, i.e. in the light of the world's structural-genetic unity.[3]

The material unity of the world can be grasped only on the basis of the structural-genetic unity of the objects of the world. This means that individual systems displaying a different qualitative level are characterized by the order governing the structural pattern, form of motion and historical age. In this way, a scale of systems can be constructed, in which a specific form of motion is assigned to a specific material system as the substratum and exponent of motion. The given system has a certain order of structural arrangement roughly corresponding to its historical age (time of origin, life span). Apart from this, systems display a certain reflection capacity, certain energy and information characteristics, spatial and temporal dimensions, etc. Real systems represent specific developmental, qualitative grades with an increasing measure of organization. Cybernetics and the theory of information (concepts of negentropy, information) opened up new possibilities for the quantification of the degree of organization. The relationship between information and energy, however, still requires a more thorough elucidation. The energy concept has a longer history and has been thoroughly elaborated in physics, while the concept of information is of a more recent date and thus awaits further research. What is absolutely indispensable for the theory of development is the concept, and/or criterion of organizational sophistication, which cannot be isolated from the quantitative aspect.

Inorganic systems (metagalaxy, galaxy, planetary systems, cosmic bodies including atoms, molecules, inorganic compounds) involve physical and chemical processes (motions) have the longest history and their level of organization and reflection capacity are low. It was in this context that living systems including man evolved: they have the shortest history, are connected with biological and psychical processes (motions) their level of organization and reflection capacity are higher. The relatively youngest systems are social systems (cultural society, contemporary society developing towards communism, hypothetical super-civilizations) which involve social forms of motion, with a high level of organization and reflection capacity, including information, homeostatic, control and auto-organizational capacity. These systems are characterized by their growth of energy and information productivity.

As far as age is concerned, the figure given for our universe is cca 18 thousand million years (beginning with the so-called singularity state within the range of 13—30 thousand million years), the Galaxy — 12 thousand million (within the range of 9—13 thousand million years), the Sun and the Earth — 4.6 thousand million years. The age of living forms ranges between 1 and 3 thousand million years. The age of primaeval man — 5 million years (range: 1—20 million years); homo sapiens — cca 100 thousand years, cultural society — cca 10 thousand years. All these are approximate orders, their mutual relation, however, does yield a certain picture. A. L. Zelmanov adduces figures characterizing the scale of object dimensions (space, time, density, interaction force). From metagalaxy up to cosmic bodies the determining force of interaction is gravitation interaction of the order 10^{-40}—10^{-12}. In macro-bodies including living organisms, molecules and atoms, electromagnetic forces prevail (order 10^{-2}). Finally, the forces operating in nuclei and kernels are of the powerful order 10^{1}. The range of spatial dimension from metagalaxies to kernels is between 10^{28} cm and 10^{-14} cm and less, temporal dimension — between 10^{18} sec and 10^{-24} sec and less, density — between 10^{-28} g cm^{-3} and 10^{15} g cm^{-3} and more.

The character of the known forms of motion and their structural substrata is indicated by data which though liable to change and requiring further precision nevertheless provide a certain criterion for comparison. According to Zelmanov, the characteristics of life and man involve a specific type of interaction as well as specific spatial-temporal intervals. Each developmental quality displays its own parameters. The theory of development cannot dispense without reliance on the theory of organization and information. At the same time, the issue of development cannot be divorced from the issue of cosmic evolution studied in Fridman's non-stationary models, the "hot universe" theory, the theory of singularity, etc. All these theories are elaborated in such minute detail that they cease to be hypotheses and offer solid ground for further study.[4]

The theory of mass and motion, elaborated by Marxist philosophy, gains further stimuli for development from the utilization of findings formulated by natural sciences such as cosmology, physics, cybernetics, biology, etc. The conception of materialist monism, which is indissolubly tied up with this theory, utilizes the idea of the structural-genetic unity of material systems relying on the theory of reflection, the theory of development, etc. The cooperation between philosophy and natural science provides the basic groundwork conducive to a further development of these methodologically indispensable ideas, their further concretization, generalization and precision.

The objective material unity of the world is reflected in an intricate manner in the unity of the scientific account of the world. Human cognition is not always fully adequate to objective unity and frequently distorts the picture by introducing simplistic and redundant elements that obstruct a correct grasp of the unity, simplicity and unequivocality of objective processes and phenomena. In its advance, science is working towards the integration of different fields of study, concepts, laws and cognition in general, towards a unified picture of the world.

The infinite variety and inexhaustibility of the objective world and the relative finiteness and restricted nature of cognition account for the fact that the attained unity of scientific accounts comprises contradictory elements, namely, that cognition is an open-ended process operating in specific segmented and seriated phases (this involves, for instance, scientific revolutions). This holds good for the contradictory aspect of theories, the aggregate of theoretical systems and scientific cognition as a whole. Unity and contradictoriness have of course their subjective as well as objective aspects. According to Gödel's theorem, each incontradictory system is based on fundamental tenets and propositions that cannot be substantiated by the means intrinsic to the system — the basic postulates of a theoretical system cannot be deduced on the basis of its logical instruments: this aim can only be accomplished through a meta-theory, i.e. a broader and more profound theory. According to G. M. Shchegolkova, incontradictoriness and completeness of a system involve what is actually a complementary relation: the system moves closer either to incontradictoriness or to completeness. In the first case, however, the system is far too empty, restricted and devoid of content, i.e. it is trivial. In the second case, the system is on the contrary excessively rich in content, range and scope and hence contradictory to a point which makes it equally uninteresting (trivial). To avoid lack of content, a system must contain a contradiction, but on the other hand must not be excessively contradictory and overdifferentiated — extreme "contentlessness" of a non-paradoxical theory is actually equivalent to the triviality of a strongly paradoxical, rich theory.

The issue concerning the construction of theories does not in a general sense involve only deductive systems. Physical theories, though they are not preeminently deductive, are fraught with similar difficulties. The problem of contradictoriness in theories is closely related to issues such as complementarity, vagueness, infiniteness, the type of contradiction, etc. J. A. Shreyder points out that invariance is a purely mathematical ideal and that increasing precision has its limits, since cognition involves the participation of the subject influencing choice (conception, mode of description, etc.). The road to a thesis is a heuristic road relying on indefinite (fuzzy) concepts; reality is infinite and inexhaustible, the subject is relatively limited and final. Hence our cognition lacks absolute precision and is indefinite.

The relevance of complementarity for an analysis of dialectical contradictoriness of both objects and their theoretical descriptions was duly stressed by a number of Soviet authors, e.g. Bazhenov, Alekseyev, etc. Briefly, this means that the measure of one contradiction is never maximum or minimum, but that contradictions are mutually related. The relation of vagueness (and hence the broader complementarity relation) specifies that the relation of opposite parameters is mathematically expressible as the constant product ($xy = \text{const}$), i.e. geometrically, as a hyperbolic curve; the product here is represented by a rectangle whose sides can have neither infinite nor zero value.

The problem of vagueness, imprecision of concepts is stressed in connection with dialectical logic, with the transitional and mediational character of opposite concepts. In Soviet literature, this aspect was pointed out by, for instance, Orudzhev and Biryukov.

Basing his argument on the numerous studies by L. A. Zadeh, Biryukov stresses the importance of sets and algorithms termed "fuzzy". These involve classes which have no sharply delimited boundaries. They can be described in terms of multivocal logic (and/or semantics), possibly also in terms of continual sets of truth values. The same applies to the equivocality of algorithms (rules). In this way we shall obtain a set ranging from deterministic algorithms to non-algorithmicity. The representation of a fuzzy set in a coordinate system is not a point representation. The set here constitutes a surface casting a "shadow" on the axis x and y (similarly as in the uncertainty relation). These problems have their relevant psychological-epistemological aspect — they are connected with the participation of the human factor. Thus in intellectual processes the dialectics of formal and non-formal, algorithmic and heuristic elements asserts its force.

Orudzhev stresses that while formal logic excludes a contradiction, dialectical logic proceeds from the premise that contradictions are implicit in the very content of thought, as they are likewise implicit in objective reality. Abstract laws of identity, contradiction and "tertium non datur" do not suffice and it is necessary to introduce concrete dialectical identity implying difference.[5]

The contradictoriness of a theory is thus connected with informational vagueness, which is accounted for by dialectical logic utilizing the complementarity principle, fuzzy sets, etc. Unity and contradictoriness are connected with the clarification of problems affecting representation systems, their levels, capacity, power, problems concerning the laws of development, the restructuring of conceptual paradigms, unification of theories, etc. In other words, this question is likewise connected with problems affecting the subjective-objective relation. Problems specific to the integration of science cannot be tackled without paying due heed to questions of objective and subjective contradictoriness.

The problem of integration of scientific cognition is closely connected with the issue of materialist monism. New concepts and approaches evolved by natural science are employed to clarify the relation between the unity of science and the unity of the world. Acting as important stimulating factors in this particular problem area are the principle of simplicity, the principle of systemicness, problems concerning the structural-genetic unity of the world, etc. The problem of unity and contradictoriness in the construction of a theoretical picture of the world is discussed with special reference to the character of dialectical contradiction, which is illustrated here in the light of recent research into complementarity, fuzzy sets, etc.

The study of problems concerning the world's material unity has a relevant bearing on the integration of scientific cognition and constitutes one of the important tasks confronting the development of scientific philosophy in the period ahead.

REFERENCES

Alekseyev, I. S.: *Kontseptsiya dopolnitelnosti.* Moscow 1978;

Ashby, W. R.: *Kybernetika,* Prague 1961;

Bazhenov, L. B. in: *Printsip dopolnitelnosti i materialisticheskaya dialektika,* Moscow 1976;

Biryukov, B. V. in: *Kibernetika i sovremennoye nauchnoye znaniye,* Moscow 1976;

Biryukov B. V. in: *Kibernetika i dialektika,* Moscow 1978;

Cailleux, A. in: *Time in Science and Philosophy,* Elsevier-Academia, Amsterdam—Prague 1971;

Goodman, N.: "Science and Simplicity. Forum Lectures", *Philosophy of Science,* Series 16.

Kaplan, S. A. (ed.): *Mimozemské civilizace,* Prague 1972;

Lenin, V. I.: *Filosofické sešity,* Prague 1954;

Orudzhev, Z. M.: *Dialektika kak sistema,* Moscow 1973;

Orudzhev, Z. M. in: *Dialekticheskoye protivorechiye,* Moscow 1979;

Petrushenko, L. A.: *Edinstvo sistemnosti, organizovanonsti i samodvizheniya,* Moscow 1975;

Polak, L. (ed.): *Variatsionnye printsipy mekhaniki,* Moscow 1959;

Prigogine, I.: *Time, Structure and Fluctuations,* Nobel Lecture, 8 December 1977;

Ruml, V.—Buhr, M.—Horstmann, H.: *Pozitivismus a vědecké poznání,* Prague 1976;

Shchegolkova, G. M.: "Paradoksalnost deduktivnykh sistem", *Vop. filos.* 1970, No 6;

Shreyder, J. A. in: *Vestnik A. N. SSSR,* 1978, No 9;

Zeldovich, J. B.—Novikov, I. Z.: *Stroyeniye i evolutsiya Vselennoi.* Moscow 1975;
Zelmanov A. L. in: *Beskonechnost i Vselennaya,* Moscow 1969;
Zeman, J.: *Teorie odrazu a kybernetika,* Prague 1978;
Zeman, J.: „Přírodovědecká racionalita a princip jednoduchosti", *Filos čas.* 1979, No 6.

1 The specific nature of organization, arrangement and structuredness has not yet been explored satisfactorily. Nor have these concepts been quantified, despite of numerous experimental projects in the field. For instance, A. Cailleux refers to the successive levels of behavioural psychism—from the medusa to man, which tries to quantify.

2 Cf. e.g. Kaplan S. A. (ed.), Mimozemské civilizace (*Extra-terrestrial Civilizations*), Prague 1972; Lem S., *Summa tekhnologii,* Moscow 1968.

3 Genetic information comprises the global programme of the structural constitution of the organism, determined by long-term reflection, while the control and communication processes of the organism, thought of short-term nature, are nevertheless notwithout their relation towards the structure of the organism.

4 Cf. Zeldovich-Novikov, *Stroyeniye i evolutsiya Vselennoi,* Moscow 1975, p. 149.

5 Dialectical logic accentuates the aspect of transition, the element of mediation. In discussing this question, Hegel, Engels and Lenin proceeded from the principle governing a body in motion, which at each point in time is and is not in the same place. The vagueness, the so-called quantity of fuzziness of motion which depends on the velocity of motion. The link between dialectical logic and the results of quantum mechanics, the relativity theory and the theory of fuzzy sets is very much in evidence. According to Orudzhev, it is incorrect to account for the difference between formal-logical and dialectic contradiction so that the first approach considers opposites in the same relation, whereas the second conception views them in the light of a different relation. The contradiction at issue is a contradiction involving the same relation (place, time), but for dialectical logic this entails not the immediacy but the mediation of contradictions and the introduction of a third, mediating link. A dialectical contradiction in which the two contrary assertions or concepts are true differs from the formal-logical contradiction in that it considers the mediating links effecting mutual interdependence of contradictions. The expression "in the same relation" refers to the extreme members of the dialectical contradiction, the qualification "in different relation" applies to the mediating link. A dialectical contradiction is the unity of opposites considered simultaneously and in the same relation (sense) — mediated by inter-links. Inter-links pertaining to the extreme members of a contradictory relation $A - B$ (non-A) possess a minimum amount of properties shared with the opposite member, and/or these properties are absent completely.

7. The Methodology of Systems Research and Mathematization

IVO ZAPLETAL

The process of penetration of mathematical and cybernetic concepts, methods findings and means into other sciences is often considered to be one of the most fundamental characteristics of contemporary science.[1] This process, in fact, is a process of the mathematization of contemporary scientific cognition. On the other hand, it is the study of complex, often multi-level phenomena and processes as systems which is considered by some other authors to be one of the fundamental characteristics of science and technology in the latter half of the twentieth century. These authors thus stress the importance and role of the systems approach (research, method) and/or the general system theory and system theories for contemporary cognition and practice.[2]

This paper aims at pointing out some logical and methodological problems and relations between these two — in our opinion — characteristic features of contemporary science, i.e. the development of system research and the mathematization of scientific cognition. This viewpoint determines both the possibilities and the limits of the paper presented.[3]

7.1 SYSTEMS RESEARCH

Systems research used for investigating one or another field of objective reality is an important instrument of cognition. At the same time, however, it cannot replace philosophy as a science on the general laws of the motion and the development of nature, society and thinking. Scientific philosophy encompasses the systems approach as one of its methodological moments but it does not reduce itself to it; its approach to objective reality is more versatile. The relatively complete cognition of a certain phenomenon in the given historical stage cannot be ensured by an individual — although significant — isolated approach in cognition, but by a totality of approaches being available. However, this conclusion is not reached within a certain individual approach, but only when we surpass it and are able to reveal the possibilities and limits of such an approach, to evaluate its cognitive potential from a broader (and higher) methodological viewpoint, and thus to find its real meaning in scientific cognition.

Modern science is a complex set of scientific theories and methods whose variety reflects the variety of the very reality investigated, as well as the variety of its motion. There exists a group of scientific methods suitable for the study of biological forms of the motion of matter, there exists a group of methods suitable for the study of social phenomena, and the like. We can see that the problem of the method of investigating (a science) cannot be isolated from the problem of the subject or object of investigation.

When evaluating the possibilities of systems methodlogy, the very conception of the methodology of science is of considerable importance as well. The Marxist methodology of science can be considered — in our opinion — to be a multi-level formation in which three levels can be specified *as a minimum* (often only two levels are distinguished): a) philosophical level, b) the level of general sciences and c) the level of special sciences. Then the following conception of the method of science (research) at all three significant levels could correspond to the levels mentioned:

a_1) method — as the most general way of analysis and systhesis in which the study of phenomena and processes being the object of science is performed. In this conception the evaluative and initial world-outlook position is included;

b_1) method — as a general means of analysis and synthesis in which the study of phenomena and processes being the object of several sciences is performed;

c_1) method — as a special way of analysis and synthesis in which the study of phenomena and processes being the object of a certain science or theory is performed.

Even here, a possible terminological differentiation arises. At the philosophical level it would be possible to speak about the methodological *principle* and/or methodological base (starting point). At the level of general sciences it would be possible to speak about an *approach* and, finally, the very special-scientific *method* would be left for the level of special sciences. It is necessary to stress the relativity of such a differentiation (principle × approach × special-scientific method) as, for instance, a certain approach can be transformed into the level of special sciences as a special-scientific method proper, and vice versa. In our opinion, it is necessary to consider from this viewpoint the problems of classifying the systems approach only at a certain methodological level. It goes without saying that the philosophical principle (method) has a unifying — both evaluative and initial — world-outlook position.

Scientific philosophy, its principles, as well as its categories fulfil the role of the methodological principle (base) in the sciences, because philosophy is — among other things — a system of requirements whose fulfilment ensures the right approach to a further investigation of reality by philosophy itself and by the special sciences as well.[4]

In this sense it is necessary to reject the efforts to raise the systems approach (method) to the level of a new "system" philosophy and/or a new "modern" world-outlook (E. Laszlo).

The relation of the methodological principle (base) to special-scientific methods is of great importance. The danger of vulgarizing this relation (if the necessary mediation is missing), particularly the relation of philosophical methods and special-scientific methods, was pointed out in a very sensible way, for instance, by P. V. Kopnin.[5]

Apart from the methodological base, it is possible to distinguish different approaches, e.g. the experimental approach or the above-mentioned systems approach, which appear in several sciences. However, systems research and mathematization of scientific cognition are again a certain concretization of the methodological potential of scientific philosophy.

The rivalry of its own kind arising sometimes in the relation of general systems theory and philosophy is not justified. As a matter of fact, philosophical categories are not only general in scope in a quantitative sense but they are also universal by their content (in a qualitative sense). The categories of the general systems theory are — with view to their content — directed only to systems objects, i.e. the objects characterized in the language of relations and/or structures, i.e. the objects on which the systems reduction of reality has already been carried out (implemented).

It is still necessary to mention the methods proper of the concretely scientific level that have been already related directly to the object of special sciences. Proceeding from the above-mentioned conception of methodology as a multilevel formation — endowed with uniformity as well, it is possible to distinguish two basic orientations in systems research (in the field of scientific cognition) — without differentiating their philosophical starting points to date.

The first orientation deals with very general problems of system research; we shall call it *general-methodological* orientation. For instance, L. von Bertalanffy's works belong to it.[6]

The other orientation approaches directly the solution of scientific problems of a less general character; it will be further called *operational-constructive* orientation. For instance, M. Mesarović's works can be included here.

As in many other classifications in the social sciences, even in this case it is not possible to demarcate a sharp boundary line between these both orientations.

If we wish to find a base on which system research (researches of systems) are established, we should probably reach the following generalization of the conception of systems research.

Systems research proceeds from the fact that phenomena and processes (elements) are understood as parts of a certain system of phenomena and processes, which is again part of some system, etc. The "initial" phenomena and processes are understood as a system as well; we sometimes speak about the system being "introduced" in them. In fact, a complex hierarchy of systems is concerned.

In L. von Bertalanffy's opinion, from the modern point of view reality appears to be an immense hierarchic order of organized entities leading to the superposition of many strata, from the physical and chemical stratum up to biological and social systems.[7]

It may, after all, be possible to characterize systems research with two basic features:

a) the phenomena and processes which are objects or subjects of research are understood as systems (requirement of a system character);

b) any phenomenon or process is part (a subsystem or even an element) of a further phenomenon or process which is again understood as a system, etc. (requirement of a hierarchic character).

(Of course, in practice one usually determines the final number of hierarchic levels.)

7.1.1 General-Methodological Orientation

Further we proceed from the existence of double possible reflection of a new method, procedure or finding. This is, above all, in the social sciences where an inadequate philosophical interpretation of special-scientific methods, procedures or findings often occurs. The phenomenon mentioned can be observed particularly among those with an insufficient philosophical base. The other reflection is the special-scientific one.

The inadequacy of philosophical reflection of a systems approach (and/or its historically younger structural method) can be observed both in the past and the present. The raising of the otherwise scientifically efficient structural method and its absolutization to structural or structuralistic ideology (or philosophy) can be given as an example. A very instructive comparison of structural method as ideology with a dialectical method can be found in the work of the Franch philosopher L. Sève.[8]

L. von Bertalanffy is among the most important representatives of the general-methodological orientation. He was originally a theoretical biologist engaged in the problems of biological growth, biological organization and structure of living organism. In the 1930s, he conceived the idea of creating a theory of open systems and later he indicated the conception of a general system theory.

L. von Bertalanffy[9] considers the general systems theory to be a logical-mathematical discipline whose object is the formulation and deduction of those principles valid for "systems" in general. L. von Bertalanffy notes the occurrence of *isomorphic laws* in different sciences. This isomorphy is said to indicate that there exist certain general systems laws which can be applied to all systems of a certain type without regard to the character of their components and relations.

However, L. von Bertalanffy reduces to a considerable extent his very demanding scientific programme at a later stage; he gives up his world-outlook aspirations and considers the general system theory as a working hypothesis.[10]

The conception of system in the individual conceptions of the systems theory is of great importance. L. von Bertalanffy understands *system as a complex of elements which are in mutual interaction.*

Apart from the concept of system the concept of structure is also included among the basic categories of system. There exists a group of scientists who identify system with structure.[11]

In what follows I shall try to indicate such a definition of basic concepts as could represent the given group of works at the given level of generality.

System can be understood as a relatively closed whole consisting of elements (parts, components) which have certain relations (linkages). The *system's structure* can be regarded as a complex of relations or linkages between the elements (parts) of systems.

The concept of *function* is usually understood in two ways. Teleologically it means activity-effect-goal, mathematically it means the kind of relation.[12]

The introduction of the concept of system's environment into the conceptual network of systems research may be the most stimulative and valuable fact from the viewpoint of methodology. It is, however, possible to find a completely erroneous interpretation of this concept with otherwise reliable authors such as A. D. Hall and E. R. Fagan. In their conception, system is understood as a complex of objects, together with relations between objects and between their properties.[13]

Objects are conceived as parts or components of a system, while these parts have an unrestricted variety.

Properties are linked to something, for instance, the properties of objects are involved here (e.g. the object-atom has, among others, the following properties of atom: number of electrons, energy of the atom, number of particles in the nucleus, atomic weight; analogically, the object-man has among others the following properties: age, a certain intellectual level, sex, a certain moral level, etc.).

The relations in which a theoretician of systems is interested are — according to A. D. Hall and R. E. Fagan — those relations which "hold the system together".[14]

Or, still better, those relations which make the concept of system useful. It is left to the person solving the problem to decide on which relations are important and which are trivial and unimportant.

According to the character of investigation and that of the problem investigated, this person will assess which relations he will deal with.

As a further concept in A. D. Hall's and R. E. Fagan's conception, the concept of system's environment is introduced: it is a complex of all objects that influence a system due to a change in their properties, as well as those objects whose properties are changed by the system's behaviour.

There is, however, an error in this definition! A. D. Hall and R. E. Fagan would involve — according to this definition — the system itself into the system's environment. Consequently, it is necessary to add to the definition mentioned that these *objects are not elements of the system*, as this is the way, and not otherwise, both authors use this concept in their further text.

The primary determination of the system's environment — analogically with the "mapping" of relations — depends to a certain extent on the person solving the problem. This rather hypothetical epistemological construction is specified and objectified in the course of further investigation.

We can define the conception of the system's environment in the following way. *The system's environment* is a complex of all elements (parts, components) influencing (acting on) the system, while not being elements (parts, components) of the system.

The system (or perhaps better said the system proper) and its environment form an area in which we are interested from the viewpoint of research.

7.1.2 Operational-Constructive Orientation

If the foregoing orientation has created conditions for applying the exact sciences and procedures to the study of complex systems (objects), then the operational-constructive orientation directly performs such precision or directly supports it to a maximum extent (for instance, by providing a precise conceptual inventory).

M. Mesarović is the most significant representative of this orientation. He constructs his mathematical systems theory (sometimes translated as the theory of mathematical systems) as a formalized discipline.[15] In one of his earlier works we can find the following definition of general systems theory:

"For a given class of sets $X = (X_1, \ldots, X_n)$ the system (abstract system) is defined as a relation to X, i.e.

$$SC(X_1 x \ldots x X_n)$$".[16]

The symbol "C" is read as "is a subset", the symbol "x" is a symbol foı the Cartesian sum, the sets $X_1, \ldots, X_n$ are called objects.

In one of his later works M. Mesarović modifies the definition of systems theory as follows: "The systems theory is a theory of formal (mathematical) models of really existing or conceptual systems."[17]

In this conception, the system is then a mathematical model of a real phenomenon. A. Rappoport comments on this statement that "a mathematical definition of system is factually a definition of a mathematical model considered as an isomorphic system which we obtain by abstracting from the real system investigated".[18]

Apart from the mathematical systems theory, which is successfully developing,[19] it is possible to range "special" systems theories or sciences (for instance cybernetics, theory of management, theory of automatic machines, theory of games, theory of graphs, and the like) to operational constructive orientation as well. After all, mathematical systems theory can also be considered one of the systems theories.

In this respect, it is necessary to keep in mind the limits of formalization and mathematization given by the dialectic of relation of the formal and the topical in methods of cognition and practice. Objective reality is a dialectical unity of the general and the individual, the general exists in fact only in individual, concrete forms and individual facts have their general substance. It goes without saying that a method making it possible to utilize the cognition of the individual for the cognition of the general, and vice versa, corresponds to the relation of the general and the individual as well.

7.2 MATHEMATIZATION OF SCIENTIFIC COGNITION

As mentioned, there are authors who consider the penetration of mathematical concepts, methods, findings and means into other sciences to be the most characteristic methodological feature of contemporary scientific cognition. The development of this process, the process of mathematization of scientific cognition, has — in our opinion — several factors. The basic factor can be seen in the specific requirements of practice. Scientific and technical progress, cosmic flights, development of means of automation as well as new forms of instruction lay ever greater claims to the development of the conceptual inventory and the means of contemporary science in general and mathematics in particular.

Automation is directly connected with the scientific and technological revolution of the present day. The very development of technology is also closely linked to the development of mathematics and its applications. "Nowadays it is already beyond doubt that it is necessary to consider a broad application of mathematical methods as one of the most fundamental characteristics of contemporary technical development."[20]

Mathematics (and/or mathematical sciences) make it possible to give precision to initial projects, for instance, in the case of technical equipment, by creating abstract models of real situations.

A further, so-called internal factor which causes such a broad utilization of mathematics in other sciences is given — in our opinion — by the character of the subject or object of mathematics and/or mathematical sciences.

It would be incorrect to consider mathematization of scientific cognition in a narrow way, only as the penetration of mathematical concepts, findings and theories into other sciences. The transfer of methods elaborated and used in mathematics (or mathematical sciences) into other fields of scientific cognition is of particularly great methodological importance. For instance, we have in mind the use of deductive methods (the axiomatic or constructive-genetic method) in their modified forms in different sciences (quantum mechanics, biology, psychology, pedagogy and economics).

In a certain sense, mathematization of scientific cognition can be ranged to a very generally conceived subject of applied mathematics which then studies and solves the problems that have arisen outside the sphere of mathematics by mathematical methods. A similar view is also expressed by G. G. Hall who states that the mathematical comprehension of reality is the goal of applied mathematics.[21]

Mathematization usually proceeds in three stages which, as a rule, follow one another.

In the first stage mathematics is used mostly for calculation purposes. For instance, the empiric laws of the course of phenomena and processes of the respective scientific discipline are expressed by means of mathematical instruments. A relatively complete description of phenomena and processes requires the transition from a qualitative description of these phenomena and processes to the study of quantitative characteristics, since any phenomenon of the real world can be understood as the unity of qualitative and quantitative aspects. The quantitative aspect exists in real processes only by means of a concrete qualitative content. Quantity as such is abstraction; in the real world quantity is always the quantity of a certain quality. It is possible to state that this type of mathematization (the first stage) is utilized by almost all the sciences, from physics to the social sciences.

The second stage (type) of mathematization of scientific cognition, which can be called the model stage of mathematization, is connected with the development and deepening of theoretical ideas and mathematical constructions in non-mathematical sciences. An informative model of phenomena and processes of a certain science is very often created; this is a model which is directly connected with formalization. This type of mathematization will be given the greatest attention further on in view of its important position in scientific cognition in general.

The third stage (type) of mathematization is one in which there the mathematical variant of the original theory already appears. Here, not only the mathematical (mostly quantitative) formulation of empiric laws or the mathematical (formal) model of a certain phenomenon or process are involved, but this is a matter of creating a new theory. These mathematical theories usually arise in fields in which there exist unambiguously characterized elementary objects.

From another viewpoint, it is possible to distinguish — as some authors do — two types of mathematization: the metrical (quantitative) and non-metrical type of mathematization. The first is connected with the above-mentioned stage of mathematization and with the utilization of classical parts of mathematical analysis; it studies in its way mostly simpler phenomena and processes admitting of numerical elaboration. Here, methods related to the theory of probability and mathematical statistics, particularly in the field of processing data obtained by observation and experiments, are of considerable importance. The instruments of mathematical analysis are often used in this type of mathematization, especially in mechanics and astronomy.

The non-metrical (non-numerical) cases (types) of mathematization belong to the second stage (model stage) of mathematization and/or overlap into the third stage of mathematization, i.e. the creation of a mathematical theory proper in a certain field. Particularly with a view to these non-numerical cases of mathematization, the efforts exerted by mathematicians to unify mathematics become ever more important. In this respect, a special position is held by N. Bourbaki with his conception of abstract mathematical structures.

7.3 ABSTRACT MATHEMATICAL STRUCTURES AND SYSTEMS RESEARCH

Contemporary mathematics is sometimes appropriately compared with a town which has older as well as newer districts. At the same time classical districts are further expanded and modernized and often visited by many people. Particularly mathematical analysis, geometry, algebra and the theory of probability belong to these older "districts". From the new parts of the town mathematical cybernetics, the theory of categories, mathematical logic, and the like, should be mentioned. The rapid development of new parts and the timely reconstruction of old parts of this "mathematical town" often lead to the question whether one town (mathematics) is still the core of the matter or whether one can already speak about several relatively independent towns (mathematical sciences). It is then quite logical that efforts are made to comprehend the unifying elements of this "mathematical town". The attempt of a group of French-speaking mathematicians who present themselves under the pseudonym of Nicolas Bourbaki belongs to the best known such endeavours. In their programme paper entitled Architecture of Mathematics the Bourbakians employed the concept of abstract structure as their starting point. It was known from Hilbert's and Dedekind's works that very extensive parts of mathematics could be developed in a logically precise and comprehensible way from a rather small number of carefully selected axioms, which led to the idea of introducing the concept of mathematical structure.[22]

The Bourbakians characterize abstract structure in the following way: for the determination of structure "several relations are usually given in which elements are found... then one postulates that the relation given or the relations given meet certain conditions (which are enumeratively calculated and hold as axioms of the structure under study). To establish an axiomatic theory of the respective structure means to deduce logical consequences from the axioms of the structure and at the same time — as far as the elements investigated are concerned — to give up any other premises (particularly all hypotheses concerning the 'nature' of these elements)."[23] Evidently, mathematical research is quite indifferent to what concrete interpretation can be given to its objects. It is simply possible to consider them as elements of abstract mathematical structures whose necessary properties are given by means of axioms.

According to the original idea of the Bourbakians different types of abstract structures are the object of mathematics. Structure is applicable to a set of elements whose nature is not determined and is given by one or several relations in which elements occur and which are determined by the axioms of the structure under study.

Of all the possible heterogeneous structures, the Bourbakians separated three basic types, the so-called "mother" structures which differ by the nature of relations they express:

Algebraic structures comply with the law of composition, i.e. the law stating that among three elements there exists a relation which defines the third element as a function of the first two elements.

In contrast to algebraic structures, the structures of arrangement do not presuppose the determination of one element by means of their elements, but they express different kinds of relation of arrangement between the elements of a set.

Topological structures (topology) — intuitively known concepts find their mathematical formulation in them: environment, limit and connection.

Further structures arise from these mother structures by means of combination (e.g. algebraic topology arises by combining algebraic and topological axioms), overlapping and different hierarchization. Even though the Bourbakians themselves admit later on that their initial hypothesis of mathematical structure has been surpassed by the concepts of category and functor,[24] they have succeeded in integrating a large part of mathematics under their axiomatically structural approach.

This is the place to review the system's conception as it is presented in the constructive orientation of systems research; thus we can clearly see its immediate relation to the conception of mathematical abstract structure as presented by N. Bourbaki. In this sense, as a matter of fact, the relation of both methodologically significant movements is quite trivial.

The problem of the reduction of classical (traditional) theory caused by the introduction of its mathematical and/or system model is considered to be the key problem of the mathematization of scientific cognition, analogically as in the case of systems research. We shall try to point out some problems of mathematization using the problems of the mathematical model and mathematical modelling. (We shall, of course, omit the calculation applications, i.e. mostly metrical applications of mathematics.) From the viewpoint of theoretical, so-called pure mathematics, there are — in principle — two possibilities of creating a mathematical model of a certain content field. On the one hand, there exists a certain "purely" mathematical theory (structure) which is then the base of a certain mathematical model. You have a situation in which — seemingly quite outside the practical needs of science or socio-historical practice — somewhere, be it in the middle or on the margin of modern mathematics, there arises a theory which (strangely enough) comes in handy for the modelling of certain content theories (or indirectly for the description of a certain part of reality). The application of the theory of groups to the problems of theoretical physics or the application of Lobachevski's geometry in the theory of relativity can serve as examples of this way of creating a mathematical model. On the other hand, there is another possibility, i.e. to create mathematical instruments (theory) directly for the needs of formalization and mathematization of a certain field.

Non-mathematical scientific disciplines lay claims to mathematics itself and a number of mathematical theories have arisen directly for the needs of mathematical modelling. The origin of the mathematical theory of games, in connection with the study of conflict situations or the origin of factor analysis in the field of psychology linked with research on human abilities, can be given as examples in this respect. This other possibility of developing a mathematical theory suitable for the modelling of non-mathematical theories conceals a problem in itself which is based on the fact that present mathematical means have arisen largely in connection with the needs of the natural sciences (above all physics) and they are, consequently, not suitable for the study of social phenomena and processes. Therefore, scientists working in marginal disciplines (mathematical biology, mathematical psychology, mathematical economics) have recently concentrated on the elaboration of mathematical instruments suitable for the study of more complex and multifarious phenomena and processes.

In connection with reflections concerning the origin of mathematical models, we have reached the formulation of the first difficulty of mathematization of scientific cognition, based on the fact that mathematicians often do not know well the field to which their theory will be applied and, vice versa, specialists in non-mathematical disciplines are not acquainted with the latest findings in mathematics. This difficulty can be removed — and has been partly removed so far — by organizing the study of mathematics in combination with non-traditional disciplines (biology, sociology, and the like). At the same time, however, disagreements arise with regard to the respective competency and status.

The second difficulty of mathematization of scientific cognition is connected with the "experiments" of mathematizing the content theory in such a way that one merely carries out a transcription of verbal statements of content theory in the symbolic form of mathematical theory (model) in order to create the semblance of scientificness. (So-called void mathematization is what is meant.) This is a certain kind of pseudo-mathematization or still better pseudo-symbolization which is actually no mathematization at all. Nevertheless, we meet with it more frequently than it would seem possible. Examples of void mathematization often occur in less developed disciplines in which their own internal content insufficiency is camouflaged with mathematical formal symbolics. In pedagogy, for instance, some authors merely denote the teacher as T and the pupil as P and their interrelation as $T \rightleftharpoons P$; thus they try to make others believe that this banal diagram is a scientific-theoretical contemplation. Naturally, only a scientific specialist to whose discipline mathematics is applied can decide on whether or not void mathematization is involved. However, there exist situations in which isomorphism between content theory and mathematical theory leads to greater vividness; of course, it can also be the other way round. Mathematics and mathematization in fact do not represent mere symbolization or quantification, but a certain abstract structuralization. (The same opinion has been expressed, for instance, by H. Lenk as well.)

The third difficulty of mathematization of scientific cognition is seen in the insufficient elaboration (stage of development) of content theory. Mathematization achieved success in all fields in which there exist unambiguously definable objects. A not very precise science cannot be changed into a precise science by any mathematization. ("A mathematized social science remains above all a social science; if it is bad, no mathematization will make a good science of it" — H. A. Simon.) The requirement concerning the elaboration of content theory and/or its conceptual inventory appears to be essential from the viewpoint of possible mathematization. After all, it is the lack of elaboration of content theory and, consequently, its inadequate mathematical model that causes the failure of many attempts to mathematize the social sciences. ("It is possible to mathematize only what has relatively precise concepts" — A. A. Markov.)

The fourth difficulty of mathematization of scientific cognition is based on the fact that there do not yet exist mathematical means able to express (model) complex phenomena and processes (e.g. social realities). Here, the systems research can be helpful in a certain sense, as a kind of pre-stage to subsequent mathematical modelling. The respective situation can be sometimes expressed so that "we have not got what we need and we do not need what we have". It can be even stated that mathematics in relation to its applications to the social sciences is at present in the period in which it was with regard to the natural sciences before the discovery of infinitesimal calculus.

Another problem which is no longer regarded as a difficulty is the problem immediately related to mathematical modelling. It is the multitude of models suitable for the description of a certain real (content) situation and/or its theory and the related problem of the selection of content theory. The mathematical theory which creates the model of a certain sphere of reality is very often a special example of a more general mathematical theory which, on the whole, no longer has such an immediate relation to reality. From the mathematical viewpoint, for instance, take the case of geometry: it turned out that it would be unreasonable to develop only those geometries which have an immediate connection with practice. The history of scientific cognition has proved that abstract non-Euclidean geometry, seemingly remote from reality and considerably more general then Euclidean geometry which models common reality, has real application in the theory of relativity. The multitude of possible mathematical theories makes the application of mathematics in non-mathematical disciplines more difficult on the one hand (the problem of selecting and adequate model), but it makes them more interesting from the heuristic viewpoint, on the other hand. The multitude of possible mathematical models is, of course, in no way suited to the conventional character of scientific cognition. On the one hand, the multitude of models expresses the fact that cognition proceeds from the cognition of different aspects of reality by means of different mathematical instruments to an ever more complete comprehension of reality; on the other hand, it also expresses the fact that mathematical means are always more flexible and interchangeable. (There exists a justification which explains the heterogeneity of mathematical models by the fact that they are only a homomorphous image of real situations on the one hand and that mathematical theories are altogether *polyvalent*, which means that they admit of different interpretations, relevant as formally logical models to a certain formalized calculus.)

Of course, in this way we come to the problems of mathematics as an ideal language (calculus) for the description of different phenomena and processes. Mathematics appears to be an "ideal technique" creating a set of concepts for other sciences. Niels Bohr who wished to underline the immense role played by mathematics in the development of theoretical cognition stated that mathematics was not only a science but also a language of science. (It might be more appropriate to speak about a calculus than a language in contemporary, formally logical terminology.)

The situation of the multitude of mathematical models acquires the following form in the light of contemplations on the language of mathematics: with regard to the plurality of mathematical languages, there exists an effort to choose in mathematical modelling such a language as is the most adequate with regard to content theory and/or to content situation, the object. It is true that a broad scale of content situations can be mathematically thematized, but scientific cognition of these situations requires an uninterrupted process of adequacy of their mathematical models.

However, it would be useful to show how concepts and methods of the theory of structures and categories facilitate the application of mathematics in other sciences.

In a number of cases it is sufficient for a scientist or a man of practice — due to the fact that the respective structures are at his disposal — to verify whether the phenomena investigated by him correspond to the axioms of one or another structure. If we meet, for instance, with phenomena which can be characterized by the properties of symmetry, it is almost certain that it will be possible to apply theoretical group structures of one or another kind in this instance. Thus all further complex and strenuous work deducing consequences from axioms disappears because it is possible to utilize at once all theorems gained during the mathematical investigation of structure. For this reason, abstract structures can be compared with finished forms which can be used while investigating and transferring information.

In spite of a number of difficulties faced by the mathematization of scientific cognition, there will be hardly anyone who can deny the practical role of mathematics. This is proved by an ever greater utilization of mathematical methods and calculation technique for the solution of topical problems of science and technology, from the organization and management of production up to such global problems as represented by the rational utilization of natural resources and the protection of living environment.

It is possible to say by way of conclusion that the system character of objects creates the necessary base for further abstractions and idealizations. After all, the mathematization of scientific cognition is ontologically based on the system conception of reality as well. By discovering the cahracter and type of system (structure) which describes a certain real object, phenomenon, process, we also determine the character and type of the corresponding mathematical description (i.e., for instance, the type of abstract mathematical structure). Using simplification, we can say that all objects, phenomena and processes correlate (in systems research) from the logical-methodological viewpoint with certain systems objects and/or structures of system and these can then be expressed by means of certain abstract mathematical systems (structures). Of course, at the same time — and vice versa — objective (content) systems objects also find their formal "pre-image" in abstract mathematical systems (structures). In this sense, mathematization of scientific cognition (in principle) in connection with systems research has *a priori* no limitations. At the same time, mathematics appears as a language able to express the character and structure of systems objects in the most adequate way.

In spite of all these facts, it is necessary to see the basic contribution of both methodological orientations, both systems research and mathematization of scientific cognition, in the sphere of the general-scientific and/or special-scientific level of methodology. It becomes evident that all aspirations of both orientations aimed at surpassing these limits to "systems" philosophy or mathematized or "mathematical" philosophy have not proved good in the history of science.

REFERENCES

1 See, for instance, Abramyan, A. O.: *Matematizatsiya znaniya* (Filozofskiy analiz), Rostov 1972.

2 See, for instance, Sadovsky, V. N.: *Osnovaniya obshchei teorii sistem*, Moscow 1974.

3 For instance, the Czech-Soviet-German collection of papers entitled *Dialektika a systémový přístup* (Dialectics and the Systems Approach), Svoboda, Prague 1979, deals with the very philosophical assessment of the possibilities and limits of systems research and mathematization of scientific cognition.

4 See Tugarinov, V. P.: "O metodakh obshchestvennykh nauk", in: *Metodologicheskoye voprosy obshchestvennykh nauk*, Leningrad 1968, p. 45.

5 See Kopnin, P. V.: *Logicheskiye osnovy nauki*, Kiev 1968, p. 235.

6 As for his writings we shall limit ourselves only to methodological considerations and leave aside their philosophical consequences whose evaluation can be found, for instance, in the works of V. A. Lektorski.

7 Bertalanffy, L. von: "An Outline of General Systems Theory", in: *The British Journal for the Philosophy of Science*, vol. I, No 2, p. 134, seq.

8 Sève, L.: "Metoda strukturalna a metoda dialekticzna", in: *Strukturalizm a marxizm*, Warsaw 1969, pp. 19–76.

9 Bertalanffy, L. von: "An Outline of General Systems Theory", in: *The British Journal for the Philosophy of Science*, vol. I, No 2, pp. 134–165.

10 Bertalanffy, L. von: "General Systems Theory. – A Critical Review", in: *General Systems*, vol. VII, pp. 1–20.

[11] We recommended speaking always about the structure of a system, not only about structure, which is what most authors of the orientation mentioned do.

[12] Analogically: Filkorn, V.: *Úvod do metodologie vied* (Introduction to the Methodology of Sciences), Bratislava 1960, p. 47.

[13] Hall, A. D., Fagan, E. R.: "Definition of System", in: *General Systems*, vol. I, I, pp. 18–28.

[14] Ibid.

[15] Mesarović, M. D.: "Osnovaniya obshchei toeiri sistem", in: *Obshchaya teoriya sistem*, Moscow 1966, p. 31, pp. 15–48.

[16] Mesarović, M. D.: "Obshchaya teoriya sistem i yeyo matematischeskiye osnovy", in: *Issledovaniya po obshchei teorii sistem*, Moscow 1969, pp. 165–180, p. 167.

[17] Mesarović, M. D.: "Teoriya sistem i biologiya. Tochka zreniya teoretika", in: *Sistemnye issledovaniya*, Moscow 1970, pp. 137–163.

[18] Rappoport, A.: "Printsip matematicheskogo izomorfisma v obshchei teorii system", in: *Systemnye issledovaniya*, 1973, p. 162.

[19] See particularly articles in the journal *Mathematical Systems Theory*.

[20] Gnedenko, B. V.: "Voprossi matematizatsii sovremennovo estestvoznaniya", in: *Materialisticheskya dialektika i metody estestvennykh nauk*, Moscow 1968, p. 175.

[21] Hall, G. G.: *The Application of Mathematical Thinking*, University of Nottingham, 1963.

[22] Dieudonné, J. A.: "The Work of Nicolas Bourbaki", in: *American Mathematics Monthly*, 1970, pp. 134–145.

[23] Bourbaki, N.: "Arkhitektura matematiki", in: *Ocherki po istorii matematiki*, Moscow 1963, p. 251.

[24] "Let us say immediately that this notion has since been superseded by that of category and functor, which includes it under a more general and convenient form. It is certain that it will be the duty of Bourbaki, who... never fears change, to incorporate the valid ideas of this theory in his works." (J. A. Dieudonné: "The Work of Nicolas Bourbaki", in: *American Mathematics Monthly*, vol. 77, 1970, p. 138).

II.
INTEGRATION PROCESSES IN SCIENCE AND THE SYSTEMS APPROACH

8. Philosophical Systems Principle, Systems Approach and General Systems Theory

V. N. SADOVSKY

8.1 INTRODUCTION

The last three or four decades have seen a rapid development of problems dealt with in systems research. Definite progress has been made in comprehending the particularities of the systems approach and in constructing a general systems theory so that hardly anyone would nowadays endeavour to deny categorically either the fact of the existence of systems research as a certain *social and scientific phenomenon* or the *special features* inherent in it which, in principle, distinguish systems research from other types and forms of *scientific knowledge*. Published sources more often than not have it that the systems approach, to use Thomas Kuhn's terminology (1970), is to be considered as a new scientific paradigm in contrast to the analytic, mechanistic, linear-causal thinking methods of classical science (Bertalanffy, 1971). Other researchers claim that certain aspects of the current scientific and technological revolution find their expression in the systems approach or the systems methods of thinking (Blauberg, Sadovsky, Yudin, 1977).

Modern systems studies represent a broad set of scientific, technological, organizational problems and the specific methods of their solution. There is a need for theoretical justification for systems studies as a rapidly developing sphere of science and technology in the second half of the XXth century and as a special form of scientific and technological activity. Such justification consists of solution of two interrelated problems, namely, the definition of philosophical foundations of systems studies and the formulation of intrascientific methodological systems conceptions. As soon as the first problem is solved, it becomes possible to establish the Weltanschauung's significance of systems research and describe their fundamental philosophical prerequisites. The solution of the second problem concerns the intrascientific theoretical justification of systems studies which should result in the construction of methodological basis for various specific systems developments.

In the second half of the sixth decade the methodological aspects of systems research began to undergo intense growth, and in the course of its nearly twenty five year history this sphere of modern methodology of scientific and technological knowledge became quite successful both from the quantitative (with respect to the range of the problems under study) and the qualitative (with respect to the results obtained) point of view. This statement can be confirmed by the conclusions of two scientometric analyses of both Soviet and Western (mainly English) literature on methodological problems of systems studies, conducted recently. The first of them performed by S. I. Doroshenko (1978) dealt with the Soviet literature on this subject that was published during 1957—1974. Just the cumulative indices of this set of publications, namely, 1074 papers and 688 authors who have written them are quite sufficient for establishing a very rapid development of research in this sphere of knowledge. The above conclusion can be made both more precise and specific if the "internal" scientometric indices of this publication set are determined.

During 1957—1974 the average increase of the set on the whole was 20 % annually, the figures for the papers being 24 % and for books — 17 %. A good approximation of such rate of growth is the exponential function. The increase was especially high in 1970—1974; during this period the set was enhanced by 679 units, whereas the preceding 13 years witnessed the increase by only 378 publications. In 1971—1974 the set was doubled, and although in the earlier period (1957—1970) the doubling of the total number of publications required a somewhat greater period of time, the mean indices of doubling of the Soviet literature publication set concerning the methodological aspects of systems research are considerably higher than the similar indices describing even the rapidly developing branches of science with the doubling period of 7—12 years. The growth of this publication set's author number is also substantially higher than average, especially during 1971—1974 when the annual increase of the number of new authors exceeded 30 %.

Another scientometric study devoted mainly to systems-methodological literature published in 1927—1976 in English (Klir, Rogers, Gesyps, 1977) has similar results. The total volume of this set was 1409 sources, while the rate of its growth is very close to that of the corresponding Soviet literature publication set.

Thus, the above scientometric data make it possible to say that at present the methodology of systems studies undergoes an extremely rapid development which is becoming "explosive" of late, not only as far as the rate of growth of the publication number, but also as far as the rate of drawing of new authors into this sphere is concerned.

8.2 SPECIALIZED SYSTEMS STUDIES AND THE METHODOLOGY OF SYSTEMS RESEARCH

Even the first publications devoted specifically to systems problems and having appeared at the end of forties and fifties (cf. Bertalanffy, 1971, ch. 2,3; Kremyansky, 1960; Lektorksy, Sadovsky, 1960) actually stressed that the field of systems studies covered both the development of *methodological principles of systems analysis* and the *specific systems research* in various areas. This statement reflected the real tendency of the systems problems evolution, the latter leading very soon to the formation of a special scientific and technological area, namely, the systems studies in the broad sense. It includes the whole set of modern scientific, technological and industrial-practical systems problems, conception, etc. Systems research in the broad sense appears as a certain aspect of the whole variety of human activity since the systems representation can be constructed with respect to its any object (and frequently the practical expediency of such "seeing" the objects has been proved).

The application of the principles of a systems study to systems studies themselves required that the analysis of the *system of systems studies* were also fixed. And as far as this factor is concerned, the identification (and the establishment of interrelations) of specialized systems studies, on the one hand, and the methodology of systems research and analysis, on the other, is of fundamental significance. Although this conclusion seems to be an obvious consequence of the relation between the methodological and the specialized scientific knowledge which is traditionally considered in philosophy, in this case it means, as a matter of fact, the determination of the main trends of systems problems development and the fixation of conditions under which these trends can be realized.

When the sphere of systems studies as a whole is divided into specialized systems studies and the methodology of systems analysis, the stress is laid upon, first, the basic difference between the problems solved in each of these fields (in the first case, it is the construction of models of specific systems and obtaining the consequences of them, in the second — the formulation of methodological principles that can aid in design and investigation of such models); second, the difference between the means of the analysis (the apparatus of specialized sciences supplemented by the methods and techniques of systems research, in the first case, and the methods of conceptual (philosophical and methodological) analysis of scientific knowledge combined with the formal apparatus of modern logic and some branches of mathematics, in the second); and, finally, third, the difference between scientific communities working out the first and the second fields (the scientific community dealing with specialized systems studies consists of specialists representing specific scientific and technological disciplines who are oriented towards systems "seeing" of the objects being studied and designed; the scientific community of systems research methodologists includes a part of the methodologists and logicians of science who are specialized in modern epistemology).

For all the peculiarities of specialized systems research and the methodology of systems analysis mentioned above, there are obvious connections between these spheres, in the sense that the second of them is intended to supply the specific means for the systems analysis of the first; no productive development of the second sphere is possible without the specific material obtained in the first sphere, etc., and even certain intersections in the set of problems and composition of the corresponding scientific communities. In spite of the presence of these common features, however, which, incidentally, is exactly the reason for unifying these two different fields into a single sphere of modern systems research, the realization of the difference between them is not only fundamental, but also necessary for the progress of both specialized systems studies and the methodology of systems research.

8.3 THE SYSTEMS APPROACH

In order to denote the set of methodological problems of systems studies, the literature, especially the Soviet philosophical one, usually employs the term "*systems approach*" (see, for instance, Blauberg, Sadowsky, Yudin, 1977). Considerable success has been achieved recently in the exposure of specific features of systems approach as a certain methodological conception. It resulted in the following basic conclusions.

In modern science and technology the systems approach appears as a special methodological concept called upon to formulate in a systematic manner the totality of methods for the analysis and design of systems belonging to various types and classes. It is important to emphasize that this interpretation of the systems approach does not imply its ability to solve the problems of general, philosophical methodology (the latter being the "internal" problem of philosophy), the methodological knowledge obtained in it plays the role of specification and further development of the corresponding branches of philosophical methodology.

In order to understand the peculiar features of the systems approach, it is necessary to adopt *the historical point of view* that regards the systems approach as a certain stage in the evolution of epistemological methods, the methods of research and design activity, the techniques for describing the nature of objects under study or subject to artificial creation. As far as its history is concerned, the systems approach is replacing the methodological conceptions of mechanicism and elementarism, its specificity and tasks being in opposition to those of these theories. Thus, for determination of the place and the functions of the systems approach in scientific knowledge the relation "mechanism — system" is basic; it fixes the difference and interaction between various ways (mechanistic and integrative) of object representation, as well as between various methods of their analysis (on the one hand, elementarist and reductionist, and on the other — synthetic, antireductionist, integrative). Accordingly, the systems approach is mainly intended to be used for the development of techniques for analysis and synthesis of objects, the description of their integrative characteristics resulting, in particular, from representation of the studied and designed objects as goal-oriented systems, the synthesis of "element" and "integrative" knowledge concerning the above objects, the investigation of the relations between the given and other systems that constitute their environment, etc.

Like any other methodological conception, the systems approach "addresses itself", first, to the corresponding specialized-scientific and technological (systems) knowledge and second, to the methodology of science on the whole, and, moreover, to philosophical methodology. These two relations are essentially different. As far as the scientists working in specific scientific and technological fields are concerned, the systems approach assumes the form of systematization of systems research methods and principles that can be used by experts in their work. The problem of employing the systems approach in specialized-scientific and technological analysis is, first of all, the problem of the corresponding experts. Besides, the systems approach which can be regarded in each given interval of time as a certain combination of methodological statements, requires development and perfection. And the working out of the systems approach can occur only in the general context of the specialized-methodological analysis, the solution of this problem belonging to the corresponding groups of philosophers and methodologists of science. Such development that was broadly reflected in systems literature of the last two decades was accompanied, on the one hand, by making the relations between philosophy and the systems approach more precise, and on the other, by identification and study of various forms of theoretical reflexion concerning systems research, namely, the general systems theory, logic and methodology of systems studies, and some other.

8.4 LEVELS OF CONTEMPORARY METHODOLOGICAL KNOWLEDGE. THE PHILOSOPHICAL SYSTEMS PRINCIPLE AND THE SYSTEMS APPROACH

The conception of various levels of contemporary methodological knowledge formulated in the end of the sixties — the beginning of the seventies by a number of Soviet philosophers (ef. Blauberg, Yudin, 1973; Lektorsky; Shvyriev,

1971; Sadovsky, 1974) played an important role in the exposure of the essence of the systems approach and in determining its connections with the dialectical-materialistic philosophy. This conception is based on the occurrence of an intense process consisting in differentiation of methodological knowledge that took place in the XXth century alongside with the rapid development of science, its structure becoming more complex, a substantial increase of the application of theoretical, abstract reasoning in it, the broad use of mathematization and formalization, etc. It would be appropriate to single out four fundamental levels in it.

1. The level of *philosophical methodology*, i.e., the analysis of general principles and the category structure of science on the whole. This area of methodology represents a branch of philosophical knowledge and is being worked out by the methods specific for philosophy.

2. The level of *general scientific methodological principles and forms of research* that includes informal general scientific conceptions (the methodology of cybernetics, methodological foundations of interdisciplinary studies, and so on), as well as formal methodological theories (the theory of hypothetical-deductive structure of science, inductivist methodology of scientific knowledge, etc.). As a matter of fact, the meaning of the general scientific property of the second level methodological conceptions is that their nature is interdisciplinary. Since the general scientific methodological conceptions do not claim the ability to solve the general philosophical problems, they are developed within non-philosophical knowledge, mainly in the frame of modern logic and methodology of science.

3. The level of *specific-scientific methodology* corresponding to the analysis of methods, principles and procedures used in specialized scientific disciplines. The main task of this methodological level is the exposure and description of the totality of methodological techniques and principles that are specific for one or another field, e.g., physics, biology, chemistry, psychology, sociology, etc. The difference between the second methodological level and the specific-scientific methodological conceptions is mainly that the latter are "tied in with the object", i.e., their statements and recommendations pertain to a strictly limited class of objects and knowledge situations that are specific for a given scientific field. It is the unified efforts of the methodologists of science and the theoreticians of the corresponding spheres of knowledge that are responsible for the construction of specific scientific methodological conceptions.

4. The level of *research methods and techniques* — the description of ways of obtaining relevant information, the experimental conditions, taking into account errors, the methods of experimental data processing, etc. At this level methodological knowledge meets methodological requirements and regulations. The latter, as a rule, are closely related to the specificity of one or another

scientific or technological discipline and are worked out in the frame of specific-scientific knowledge.

From the point of view of this classification the systems approach belongs *to the second level of methodological knowledge.* A detailed justification of this statement can be found in Blauberg and Yudin (1973); its result was not only the negative characteristic of the systems approach (as a non-philosophical methodology), but also the positive one (as general scientific, interdisciplinary methodological knowledge). According to the conception of methodological levels, the philosophical methodology (dialectics) is basic for any forms and types of methodological knowledge, and each more general methodological level plays a determining role with respect to the levels of methodology dealing with less generalized statements. This served as the foundation for the natural explanation of both the connection between "dialectics — systems approach" (the connection between philosophical methodology and general scientific non-philosophical methodological conception) and the relative independence of systems approach, combined with its role in the formulation of specific scientific methodological and systems principles and statements.

It seems to us that the above interpretation of the systems approach status, as well as its interaction with dialectics, takes into account sufficiently adequately also the modern level of theoretical realization of the essence and specificity characterizing the methods of systems studies and the assertion of the primacy of philosophical methodology with respect to other levels of methodological knowledge, which is fundamental for Marxist philosophy. This interpretation was quite broadly reflected in the Soviet methodological literature.

At the same time, the recent Soviet systems literature contains in addition a different treatment of systems approach. L. Naumenko, for example, regards the systems property as "the most important feature of dialectical method" (Naumenko, 1974, p. 96). In his analysis of the role of the systems approach in Hegel's dialectics he arrived at a conclusion that "the whole totality of categories in Hegelian logic is nothing but the category structure of the systems approach" (Naumenko, 1974, p. 98). Thus, in this interpretation the systems approach appears as an essential feature or aspect of dialectics.

It seems as if we are dealing with two considerably different conceptions of the systems approach, and in this connection it is necessary to clarify to what extent each of them is valid. I think, however, that the real difference between these two conceptions is terminological rather than meaningful. To demonstrate this, the *philosophical systems principle* must be distinguished from the systems approach. The first of them "means that a phenomenon of objective reality regarded from the point of view of the systems whole regularities and the interaction of its components forms a special epistemological prism, or a special "dimension of reality" (Kuzmin, 1976, p. 10). As can be shown (which is done convincigly in Kuzmin, 1976; Naumenko, 1974; Ogurtsov,

1974), the systems principle interpreted in this way and including the philosophical ideas of the systems connectedness between the elements of objects, wholeness, structural property etc. was important in the history of development of all philosophy and became best expressed in the Marxist-Leninist philosophical conception. One would be therefore completely justified to say "Marxist theory and methodology contain the systems principle as one of their most essential components" (Kuzmin, 1976, p. 20). I would like to emphasize the fact that the point in question here is exactly the philosophical systems principle, and not the systems approach. As for the latter, its treatment as the general scientific non-philosophical methodological conception, presented, in particular, in this paper, seems to be the most adequate, both from historical (the term "systems approach" itself was introduced literally several years ago) and meaningful (the systems approach principles do not claim philosophical generality) points of view. I would say in addition that the notion "approach" (in the sense of a research approach) is a non-philosophical one; philosophy usually deals with "methods", "fundamental principles of research", and so on, rather than with "approaches". This is another reason why the term "systems approach" should not be transferred, with the radical change of its contents, into the sphere of philosophical methodology, where this concept was never used in the course of history and where there are quite adequate terms, such as the "systems principle", the "systems method" for denoting the corresponding field.

8.5 GENERAL SYSTEMS THEORY

Systems thinking in the form of attempts to construct a *general systems theory* involves an essential gnoseological difficulty. In numerous attempts at a construction of a general systems theory put forward in contemporary philosophical and scientific works, this theory is equally interpreted as a *generalized scientific theory* (general scientific and technological theory of systems) and as a *certain methodological conception* (general systems theory as a methodology of investigating systems objects). Often, the specific problems of scientific generalization and methodological research are applied to the same version of constructing a general systems theory. Such synchretism manifests an essential theoretical uncertainty about the objectives and tasks of a general systems theory.

The term 'general systems theory' accounts, to a certain extent, for this theoretical uncertainty. The English use of this term ('General Systems Theory') allows for its intepretation both as a General Theory of Systems and a Theory of General (generalized, abstract) Systems. It is true, the corresponding German term 'allgemeine Systemtheorie (Systemlehre)' which, incidentally, appeared before the English term (Bertalanffy, 1949), can be interpreted only in one sense as 'a General Theory of Systems', however, the two English meanings of the term have had an effect on the development of the theory.

When one tries to realize theoretically the true meaning of the general systems theory interpreted only as a general theory of systems objects, one generally emphasizes an 'unusual difficulty' in selecting a proper level of generality of such a theory, since quite a general theory might easily turn out to be trivial in content, while a substantial narrowing of its scope of application will have to make us give up its interpretation as a general theory. M. Mesarović who discusses the problem in detail and who considers the general systems theory both 'as a foundation of other specific theories' and as encompassing 'different types of already existing specific theories' (Mesarović, 1964, pp. 1—3 and elsewhere) believes that the problem of the proper level of generality of the general systems theory can be solved if we interpret a system as a relation (Mesarović, 1964, pp. 5—8). As a result, the general systems theory turns out to be a fragment of the theory of relations, or, to be more precise, it is constructed on the basis of the theory of relations. However, though the generality of the term 'relation' is rather high, the general systems theory constructed on the basis of the theory of relations is far from being universal and actually appears in M. Mesarović's writings as a theory of specific classes of abstract systems.

On the other hand, the task of constructing a theory of abstract systems, i.e. the realization of the general systems theory in the latter sense, envisages, in the long run, the construction of a *general system* of abstract systems, since it should by nature comprise a theoretical description of the multiplicity of abstract systems. And the problem of the level of generality and the meaningfulness of the theory being constructed arises again in this case.

The difficulties mentioned above, in the final analysis, affect the solution of the problem of whether it is possible to construct a general systems theory as a general (generalized) scientific and technological theory. We have just pointed out the difficulties in realizing this task which are connected with the selection of the proper level of generality and the discovery of a non-trivial meaning of such a theory. Now we should like to bring out another point.

When we pose the problem of constructing a general systems theory as a generalized (sometimes it is referred to as universal) scientific and technical systems theory, we presume that such a theory should be constructed on the basis of a special systems logic and methodology. However, this section of systems knowledge should be included in the general systems theory, since the latter claims to be the *general theory of systems*, and therefore, it is impossible to construct a general systems theory as a generalized scientific and technological systems theory alone. Thus, the general systems theory as implied by the literal meaning of the term should be at the same time both a general scientific and technological systems theory and a general theory of logic and methodology of systems research, thus evidently depriving it of the status of a scientific theory proper. It should also be pointed out that the construction of a general systems theory as a conception which combines problems of the general scientific and technological systems theory and the general theory of logic and methodology of systems research presupposes the solution of the problems of selecting the proper level of generality of such a theory and determining the ways of obtaining a non-trivial meaning in it, which, as has been shown, is far from being easy to accomplish.

It becomes clear from the discussion that the predominant interpretation of a general systems theory as a theory combining the tasks of scientific and technical generalization of systems knowledge and the problems of logic and methodology of systems research in fact contains serious difficulties and contradictions.

The search for the solution to the problem of specifying the actual meaning of a general systems theory has prompted us to think that it would be expedient to interpret this theory as a metatheory of some sort in relation to the specialized (biological, psychological, sociological, technological, etc.) systems theories and various systems conceptions and constructions. In doing so, we proceed from the following reasons. There can be no doubt about the scientific foundation of various special or specialized systems theories (as applied both to different branches of knowledge, i.e. the theory of biological, social, etc. systems, the theory of large technological systems, and to various aspects of systems research, i.e. the theory of static and dynamic systems, the theory of open systems, the theory of hierarchical systems, etc.). Being limited in subject-matter and the means used, such theories, naturally, do not claim to be universal (cf. Sadovsky, 1972). However, while constructing them we have to make use of the generalized ideas of systems. Such ideas are borrowed from the general philosophical conceptions of systems research, from relevant sections of logic and methodology of science dealing with the description of the methodological and logical procedures of systems analysis. However, this knowledge does not fully satisfy the needs of special systems theories. The understanding of the general principles of constructing knowledge about systems objects is of significant importance for them, and this can be provided by the theory of systems theories, i.e. *the metatheory of systems research.*

During the twentieth century it has been realized that the problem of generalizing scientific knowledge can be solved not only by constructing a generalized theory in the strict sense of the word (generalization in relation to some particular knowledge or certain classes of such particular knowledge), but also as a result of creating a metatheory which analyzes the structure, means of expression and construction methods of a certain theory or a certain class of theories. A metatheory is directly concerned with a set of specific theories providing generalized conceptions for appropriate knowledge and thus generalizing the subject-matter content expressed in this knowledge.

Thus, the general systems theory as a metatheory being concerned with a class of specific systems theories formulated knowledge related to the whole class of special systems theories. Such knowledge, therefore, represents the *general systems knowledge*. Besides, since the metatheory should satisfy the conventional principles of theoretical construction, the assetions being formulated in the framework of the systems metatheory should be arranged in a *strict theoretical form*. As a result, by realizing the programme of constructing a general systems theory as a metatheory we, on the one hand, fulfil the objective of constructing a general systems theory as a general theory of systems, and, on the other hand, remove all the difficulties and paradoxes which arise while interpreting the general systems theory only as a generalized scientific and technological systems theory. The investigation of specific systems with an appropriate degree of generalization is the business of special systems theories, whereas the task of the general systems theory as general and a theory consists in analysing systems theories, and only through them it is related with the world of real systems.

In the present article it is not our task to carry out a full-scale construction of a general systems theory as a metatheory, quite apart from the fact that this would be exceedingly complex. Therefore, in concluding this section of the article we shall dwell on some general principles of constructing a systems metatheory (further considerations to this effect can be found in Sadovsky, 1972, 1974).

At present the technique of metatheoretical research has been evolved mainly for the metatheoretical analysis of formal scientific disciplines, mathematics and logic. Object languages of mathematical and logical theories lend themselves to a strict formalization, the extent of the latter being such that a Turing machine can be built over a set of propositions of appropriate object languages enabling operations to be performed over these expressions in a more formal way irrespective of the content expressed by them. As a result, the corresponding metatheories — metamathematics and metalogics — can be constructed in a purely formal manner.

The situation differs in the case of the metatheoretical analysis of empirical knowledge to which the specialized systems theories also belong. The formalization means nowadays enable us to formalize scientific knowledge only as regards certain aspects; therefore the systems theories in our analysis should be considered as not completely formalized. It suffices that such theories are expressed in relatively strict languages and the transformation of some expressions into others is intuitively apparent in them. The systems meatheory should be developed in this connection as a *non-formalized metatheory*. Such a programme of developing the general systems theory as a metatheory is, however, in complete accord with the practice of metamathematical and metalogical investigations in which the corresponding metatheories were subjected to formalization only after the meaningful construction had been realized.

Another important principle of constructing a systems metatheory consists in determining the *specific qualities of the metalanguage* in terms of which such a metatheory can be developed. Generally, the metalanguage is to meet two basic requirements. Firstly, it must contain the object language as its part, i.e. all the propositions of the object language should be transformable into the propositions of the metalanguage; and secondly, it should include a certain set of specific terms and propositions (metaterms and metapropositions) which have no corresponding analogues in the object language and which are used to describe the construction, structure, logical properties, etc. of the object language. Thus the systems metalanguage the construction of which demand that these requirements should be met must contain not only a definite set of general metatheoretical terms and propositions ('truth', 'meaning', 'sense', 'non-contradiction', etc.) but a corresponding set of *specific systems metaterms* and *metapropositions*, enabling various object systems theories to be compared, the invariant content of the systems terms 'system', 'structure', 'wholeness', 'connection', 'hierarchy', etc. to be singled out, as well as the interrelations between the specific systems conceptions and constructions in science and technology and various versions of systems theory (with a varying degree of generality) to be established.

The general framework of the systems metatheory (peculiarities of the assertions formulated within its scope and singling out the main branches of such a theory) depends, to a large extent, on the initial assumptions adopted in it and the definitions of the principal terms of systems research being developed. The metatheoretical analysis of the terms 'system', 'structure', 'wholeness', etc. differs considerably from the methods of defining these terms in the specialized systems theories. In the latter case use is generally made of the conventional meaningful intuitive considerations; for example, a system is being considered as a whole set of interconnected components, and these conceptions are being specified depending on a particular scope of objects under investigation. As a matter of fact, the only reason for one or the other definition of the terms 'system', 'structure', etc. in specific systems theories is the more or less strict introduction of these terms into special languages of appropriate systems theories and the possibility of their operative use. Naturally, the resultant knowledge can in no way claim the status of the general systems knowledge.

In defining the basic systems terms in the framework of the systems metatheory it is necessary, in addition to satisfying the above mentioned conditions, to compare the intuitive content of appropriate terms in their various scientific and technical uses with the content of the terms which are being introduced and to carry out a *comparative analysis* of various approaches to the definition of these terms. As a result, it becomes possible to create the necessary conditions of defining the principal systems terms and to systematize their various aspects which enables us to describe the scope of the systems metatheory. In the next section of the article treating the metatheoretical analysis of the term 'system' as an example, we shall show how such an investigation can be carried out in practice.

8.6 RELATION, SET, SYSTEM

In numerous specialized systems theories, as well as in certain versions of a general systems theory (for example, in M. Mesarović's conception of a general systems theory) the system is defined as a relation. M. Mesarović, in particular, gives the following definition: "The system, S, is the relation defined on the sets $V_1, \ldots, V_n$, i.e. S is a (proper) subset of the Cartesian product of the sets $V_1, \ldots, V_n$, $S \subset V_1 \times \ldots \times V_n$', where $V_1, \ldots, V_n$ are the sets of values of attributes of the objects under investigation" (Mesarović, 1968, p. 62).

Widely known is the understanding of a system as a set (of elements, components, parts, actions, operations, functions, etc.) which is subjected to certain restrictions. A. D. Hall and R. E. Fagen, for example, define a system as "a set of objects together with relationships between the objects and between their attributes" (Hall, Fagen, 1956, p. 62). O. Lange treats a system as "a set of coupled active elements" (Lange, 1965, p. 17).

There is no need to overestimate the differences in the interpretation of a system as a relation or a set. The relation itself in the framework of modern mathematics is defined as a certain set, as a subset of the Cartesian product of the basic sets. And therefore, the only difference between these two definitions of the term "system" consists in whether or not the basic set itself is included into the system, with the relations, connections, etc. being introduced into this set. Therefore, we can speak about a single set theoretical interpretation of the term "system", which has at present probably gained the widest recognition.

However, we know serious objections to such an interpretation of the term "system". As far back as in the early 40's A. Angyal (1969) formulated a number of such objections. In his opinion "the structure of the wholes cannot be described in terms of relationships" (Angyal, 1969, p. 17) and "a system is not a complex relation" (Angyal, 1969, p. 18). A. Angyal gives, in particular, the following arguments to prove the proposed assertions. While relations are binary or can be reduced to being binary, a system, being a whole organization and comprising an indefinite number of elements, cannot be reduced to binary relations and, therefore has a completely different logical status (Angyal, 1969, pp. 18—20). The immanent properties of correlated objects form a basis for relations; whereas the object participates in the system on the basis of its positional value in the system (Angyal, 1969, pp. 20—21). While in a relation the connection between the correlated objects is direct and without intermediate links (object a is directly correlated to object b, and b, to a), the elements of a system are interconnected only through their connection with the whole (Angyal, 1969, pp. 21—24). As a result, A. Angyal arrives at the conclusion that relations and systems are two different logical genera and, in any case, systems cannot be deduced from relations.

Comparatively recently E. R. Rannap and Yu. A. Schreider have voiced serious methodological objections also to the interpretation of a system as a set (Rannap, 1971, Schreider, 1971). In forming a set the elements are primary, and definite groups of these elements form the specific sets. The primary characteristic of the system is that it is some "whole composed of interacting (connected) parts" (Schreider, 1971, p. 5). The elements for the system are not given in advance; they are constructed (or selected) in the course of dividing a system, each system lending itself to various divisions. Each division of the system is a set but the system itself is not a set (Schreider, 1971, p. 5). "A system is not a set but it might be considered as a set" (Rannap, 1971, p. 5). Refering the set theoretical interpretation of a system to the "second level of its investigation when we have already established the method of division into elements for a specific purpose of investigation" (Rannap, 1971, p. 5) is an immediate conclusion drawn from such an approach to the understanding of a system.

In accord with these considerations E. R. Rannap and Yu. A. Schreider arrive at the definition of a system as a class of sets $S = \{M_s^i\}$, where $i = \alpha$, β, $\gamma, \ldots$, a many-to-many correspondence $\Phi_{\alpha\beta} : M_s^\alpha \rightarrow M_s^\beta$ being established for each pair of sets (Rannap, 1971, p. 5). Such a correspondence makes it possible to interconnect various descriptions (divisions) of a system. We consider such an approach to be an essential step forward on the way of overcoming the limitations of interpreting a system as a set and a relation. In terms of a systems metatheory this approach can be described in the following way.

The essential characteristics of a systems investigation, viz. *the principles of wholeness and hierarchicality,* are the basis for a metatheoretical analysis of the term 'system'. Derived from them, in particular, is the primacy of a system as a whole over its elements and the fundamental hierarchical organization of any system. To these principles we should add *the principle of plurality in describing any system,* since in order to obtain adequate knowledge of a system it is necessary to construct a *certain class of its descriptions,* each of them being capable of covering only the definite aspects of wholeness and hierarchicality of a given system. This principle, as matter of fact, is a corollary to *the fundamental relativity* of any description of a system, and thus an expression of the paradoxicality of systems thinking and the existing means of eliminating the paradoxicality.

In general we can ascertain that each system under investigation requires three different levels of its description: (1) from the point of view of its external wholeness properties inherent in it; (2) from the point of view of its internal structure and the 'contribution' of its components to the formation of the wholeness properties of a system; (3) from the point of view of understanding a given system as a subsystem of a larger system. However, in scientific and technological practice the number of levels of describing a system is generally larger. Not only can any of the named levels of describing a system be differentiated (for example, in analysing the internal structure of a system we can probe more deeply by subjecting to further division those elements of the system which in another description are assumed to be not further divisible, etc.), but various descriptions of systems can also be constructed for any given level taking into account the fundamental relativity of any system description.

Thus, in a class of sets with the aid of which a system is defined it is necessary to distinguish: a subclass of sets $\{M_s^i\}$, where $i = \alpha, \beta, \gamma, \ldots$ (these sets are the results of divisions of the basic object of a system into its elements), a subclass of sets $\{L_s^j\}$, where $j = a, b, c \ldots$ (these sets are formed as a result of further divisions of elements of the basic object or a system) and, finally, a subclass of sets $\{K_s^k\}$, where $k = 1, 2, 3, \ldots$ (the object-system under investigation participates as a certain element in every one of these sets). As a result, we arrive at the following general definition of the term 'system'. A system S is a class of sets $S = \{M_s^i, L_s^j, K_s^k\}$, where $\{M_s^i\}$, $\{L_s^j\}$ and $\{K_s^k\}$ are different sub-classes having a number of correspondences for each pair of sets taken both from one sub-class and from different sub-classes.

The proposed understanding of a system realizes all the main functions which a metatheoretical definition of this term should perform. It is in accord with the principal points of the intuitive interpretation of a system, since it determines the plurality of system elements and the connections between them. This definition allows a comparative analysis of the special definitions of a system which results, in particular, in various versions of the set theoretical interpretation of a system being considered as derivative, since it depends on the basic understanding of a system as an hierarchically organized wholeness. From such a basic understanding it is possible to construct a whole variety of specialized interpretations of a system.

8.7 PRESSING QUESTION IN THE DEVELOPMENT OF GENERAL SYSTEMS METHODOLOGY

We have considered only several, most essential, as it seems, to us, problems of systems methodology and have attempted to summarize the results of their working out. Certainly, many other, more special questions of methodology of systems studies the analysis of which occupies a prominent place in the corresponding literature remained beyond the framework of our approach. Since it is not possible, because of the limited size of this paper, to treat these topics, we shall formulate in conclusion a number of the most important problems of methodology of systems research which, as far as it appears to us, require at present their solution (see also Sadovsky, 1980).

It would be appropriate to include into this list the following problems.

1. Making the definitions more precise and constructing formalized descriptions of the basic concepts of systems research methodology. The point in question here is such concepts as "system", "element", "structure", "wholeness", "connection", "hierarchy", "system-environment relation", etc. Very considerable work has been performed in this area (see for example Klir, 1969; Uyomov, 1978; Urmantsev, 1974), and this is quite natural, since any attempt at development of the systems approach or construction of the general systems theory sooner or later necessarily faces this problem. In spite of this, however, now there is no system of the basic concept of systems studies which is more or less generally accepted. We have only a number of various approaches to the construction of such system. As a rule, they are quite isolated from each other, the same systems concepts are defined in essentially different ways, the definitions are not rigorous, etc. Because of all this, the above problem becomes very pressing.

2. The theoretical description of specific methods of systems research that distinguish these from other types and forms of scientific study (substrate, genetic, functional, structural, etc.). In spite of the fact that the formulation of this problem itself is certainly not new, it seems to us that no serious attempts of its solution have been undertaken up to now. For this reason, at present, as far as their methodology is concerned, systems studies appear as an agglomeration of a variety of most diversified methods, procedures and techniques with respect to which neither their special systems content nor their place and functions in the system of systems research methods have been clarified. To-day the latter does not exist at all; this is why the development of the above problem is necessary.

3. The analysis of epistemologically peculiar features of systems research. The urgent need of the working out of this problem is explained by the fact that as the object as a system is studied or designed, many classical epistemological topics, such as the relation between the object and the subject in knowledge, the connection between the truth and correctness of the decision made by the subject, etc. acquire a new meaning. This is mainly due, first, to the high degree of complexity of typical systems problems the solution of which, as a rule, is only plausible; second, to the more important (as compared to the classical scientific studies) role of the subjective factor in the working out of systems problems; and, finally, third, to the process of systems investigation being internally contradictory; the latter is expressed, for example, in the presence of systems thinking paradoxes (cf. Sadovsky, 1974, pp. 232—246).

4. *The comparative analysis of conceptual structures and the fundamental principles of the systems approach (the general systems theory) and the set-theoretic methodological conception.* This task is closely related to problems 1, 2, 3 and can be regarded in a certain sense as their concretization. The reason why I single it out as a special problem is that many aspects of the opposition "systems — set-theoretic research" are decisive both for the understanding of the essence of the systems approach and for constructing the specific systems conceptual apparatus. Indeed, the classical science of the XIXth — the first half of the XXth century is based on set-theoretic abstractions and principles, while science and technology of the second half of the XXth century which have acquired the methods of systems research make an attempt to overcome the limited nature of set-theoretic methodology. This widely recognized intuitive opposition, however, must be expressed in a systematic and rigorous way. The first steps in this direction have been undertaken with Schreider (1978).

5. *The construction of the general systems theory as a systems metatheory.* The significanée of the development of this methodological trend has been discussed with sufficient details earlier.

6. *The creation of the theory of systems classification.* The nature of such theory can be only methodological, while its results are equally necessary both for the development of systems research methodology and for specialized scientific analysis of specific (material as well as ideal) systems.

7. *The working out of methodological foundations of the simplification theory.* Since the systems the study of which is interesting from theoretical and from practical point of view are extremely complex (the living organism, the human brain, the social organizations, the industrial conglomerates, the control systems, etc.), the only possible strategy of their study and/or design is the operation with the simplified models of such systems. Although very many systems theoreticians mention the importance of formulating the theory of systems simplification, no substantial results have been obtained in this field as yet.

8. *The development of methodological foundations of the theory of hierarchical systems.* Recently this part of the general systems theory has been drawing greater and greater amount of attention. The works of M. Mesarović and his

co-authors (1970) which expose a number of essential characteristics of any hierarchical organization made a certain contribution into the construction of the theory of hierarchical, multi-level systems. Further development of this field is slowed down substantially by the lack of sufficient methodological foundations for such theory.

9. The study of methodological problems of systems analysis. At present the systems analysis is the most widely used means of systems research for solving practical problems. It is obvious that the efficiency of this means depends mainly on how profoundly its methodological nature is understood. Nappelbaum (1980) suggested an interesting program for the working out of methodological and theoretical problems of systems analysis.

10. The construction of methodology of science based on the principles of the systems approach and the generel systems theory. No matter how paradoxical it may be, the systems approach, representing one of the scientific and methodological trends in modern science, has exercised up to now practically no influence on the construction of science methodology. And the latter (it must be emphasized that the point in question is not the philosophical methodology but the specialized scientific and methodological conceptions) undergoes now a period of serious restructuring which is expressed, first of all, in the transition from the hypotheticaldeductive justification of the scientific knowledge ideas to the historical and methodological models of the structure and development of science. Such models, e.g., Kuhn's paradigma (1970), I. Lakatos' methodology of research programmes (1970), a number of science evolution models suggested recently, employ, explicitly or implicitly, the systems approach principles both for the understanding of their objects (a scientific theory, a historic sequence of scientific theories, scientific community, etc.) and for construction of a theoretic description of their "activity" (the fixation of the wholeness of scientific knowledge, the analysis of the relation between philosophy and specialized scientific knowledge, establishing systems connection between the empirical and theoretical levels of knowledge, etc.). The problem is to turn these systems intentions that are scattered and not always clearly formulated in the interpretation of the methodology of science into a rigorous scientific theoretical construction.

As it seems to me, such are the main tasks of the systems research methodology, the development of which is now of the greatest interest. In conclusion I would like to stress the fact that I have been dealing here only with the problems of system study methodology, rather than with the pressing questions of systems studies on the whole. Certainly, the latter do not coincide with list given above.

REFERENCES

Angyal, A., 1969, "A Logic of Systems", in Emery, F. E. (ed.) *Systems Thinking. Selected Readings*, Penguin Books: 17–29.

Bertalanffy, L. von, 1949, "Zu einer allgemeinen Systemlehre", *Biologia Generalis* 19: 114–129.

Bertalanffy, L. von, 1962, "General System Theory—A Critical Review", *General Systems*, 1–20.

Bertalanffy, L. von, 1971, *General System Theory: Foundations, Development, Applications*, London: Allen Lane the Penguin Press.

Blauberg, I. V., Yudin, E. G., 1973, *Emergence and Essence of the Systems Approach*, Moscow: Nauka Publishers (in Russian).

Blauberg, I. V., Sadovsky, V. N., Yudin, E. G. 1977 *Systems Theory, Philosophical and Methodological Problems*. Moscow; Progress Publishers.

Caws, P., 1968, "Science and System. On the Unity and Diversity of Scientific Theory", *General Systems* 13: 3–12.

Doroshenko, S. I., 1978, "Scientometric Data of the Soviet Literature on Systems Research", *Systems Research, Yearbook 1978*: 127–135 (in Russian).

Hall, A. D., Fagen, R. E., 1956, "Definition of System", *General Systems* 1: 18–28.

Klir, G. J., 1969, *An Approach to General Systems Theory*, New York: Van Nostrand Reinhold Company.

Klir, G. J., Rogers, G. in coop. with Gesyps, R. G. (eds.), 1977, *Basic and Applied Systems Ressearch: A Bibliography*, State University of New York at Binghamton, Binghamton, N. Y.

Kremyansky, V. I., 1960, "Certain Peculiarities of Organisms as System from the Point of View of Physics, Cybernetics and Biology", *General Systems* 5: 221–230.

Kuhn, T. S., 1970, *The Structure of Scientific Revolutions*, Chicago: University of Chicago Press.

Kuzmin, V. P., 1976, *Systems Principle in Marx's Theory and Methodology*, Moscow; Politizdat (in Russian).

Lakatos, I., 1970, "Falsification and the Methodology of Scientific Research Programmes", in I. Lakatos and A. Musgrave (eds.): *Criticism and the Growth of Knowledge*, Cambridge, Cambridge University Press.

Lange, O. 1965, *Wholes and Parts. A General Theory of System Behaviour*, Oxford, Warszawa: Pergamon, Polish Scientific Publishers.

Lektorsky, V. A., Sadovsky, V. N., 1960, "On the Principles of Systems Research (Related to L. Bertalanffy's "General System Theory"), *General Systems* 5: 171–179.

Lektorsky, V. A., Shvyryov, V. S., 1971, "Urgent Philosophical and Methodological Problems in the Systems Approach", *Voprossi filosofii*, No 1: 146–153 (in Russian).

Mesarović, M., 1964, "Foundations for a General Systems Theory", in: Mesarović, M. (ed.): *Views on General Systems Theory*, New York: John Wiley and Sons: 1–24.

Mesarović, M., 1968, "Systems Theory and Biology—View of a Theoretician", in: Mesarović, M. (ed.): *Systems Theory and Biology*, Berlin, Heidelberg, New York: Springer Verlag: 159–87.

Mesarović, M., Macko, D., Takahara, Y., 1970, *Theory of Hierarchical, Multilevel Systems*, New York: Academic Press.

Nappelbaum, E. L., 1980, "Systems Analysis as a Development Programm—Structure and Main Topics", *Systems Research. Methodological Problems. Yearbook 1979:* 55–79 (in Russian).

Naumenko, L. K., 1974, "Hegel's Dialectics and the Systems Approach", *Filosofskiye Nauki,* No 4: 95—103 (in Russian).

Ogurtsov, A. P., 1974, "The Development of the Systemity of the Scientific Knowledge", *Systems Research, Yearbook* 1974: 154—186 (in Russian).

Rannap, E. R., 1971, "Systems Analysis of Invention Descriptions, *Nauchno-Teknicheskaya Informatsiva.* Serial 2: *Informatsionnya Protsessy i Sistemy* No 6: 5—10 (in Russian).

Sadovsky, V. N., 1972, "General Systems Theory: its Tasks and Methods of Construction", *General Systems,* 17: 171—179.

Sadovsky, V. N., 1974, *Foundations of the General Systems Theory. A Logical and Methodological Analysis,* Moscow: Nauka Publishers (in Russian).

Sadovsky, V. N., 1980, "Systems Approach and General Systems Theory—State of the Art, Main Problems and Development Perspectives", *Systems Research. Methodological Problems. Yearbook 1979:* 29—54 (in Russian).

Shreider, Yu. A., 1971, "On the Definition of a System", *Nauchno-Teknicheskaya Informatsiya, Serial 2: Informatsionnye Protsessy i Sistemy* No 7: 5—8 (in Russian).

Shreider, Yu. A., 1978, "Set Theory and Systems Theory". *Systems Research. Yearbook 1978:* 70—85 (in Russian).

Uyemov, A. I., 1978, *Systems Approach and General Systems Theory,* Moscow: Mysl Publishers (in Russian).

Urmantsev, Y. A., 1974, *The Symmetry of Nature and the Nature of Symmetry. Philosophical and Scientific Aspects,* Moscow; Mysl Publishers (in Russian).

9. The Dialectics of the Development of Science and the Systems Approach

ZDENĚK JAVŮREK

The current development of science has been characterized by a number of new phenomena and approaches, methodological procedures and theories which — in the process of studying reality — assume almost a conceptual character.

The growth of the role of theories, approaches, methods and corresponding categories and concepts, which are frequently — unlike the individual special science theories and methods, approaches and categories on the one hand and philosophical theories, methods, approaches and categories on the other — referred to as belonging to general science, has now become more than obvious. It is sufficient to point to such theories and approaches as cybernetics, the theory of information, the theory of games, modelling, or the systems approach, etc. In this context also the growth of the role of certain concepts and categories, such as e.g. feedback, information, model, function, system, structure, element, etc., has been going on.

The mounting role of these theories, approaches, categories, and corresponding methods has been accompanied by far-reaching changes related to the rapid development of scientific learning as such and at the same time to an enormous expansion and promotion of science's social functions, its role in the complex process of cognition of intricate problems that emerge before mankind and then assume the dimension of global problems, i.e. issues of paramount importance for the rate and history of mankind.

In this connexion, Marxist-Leninist philosophy frequently refers to various levels or degrees of contemporary science methodology.

The growth of the significance of such theories and approaches as cybernetics, theory of information, modelling, the systems approach, etc. has for years been in the foreground of attention of philosophers who proceed from the theoretical as well as methodological positions of Marxism-Leninism. This is in harmony with the teaching of Karl Marx, Friedrich Engels and Vladimir Ilyich Lenin who never regarded the philosophy of dialectical and historical materialism as a rigid, theoretically and methodologically closed dogma but rather as a living theory and method which develops in connexion with the theoretical and practical tasks posed by everyday life. They regarded materialist dialectics, this revolutionary core of Marxism, even in relation to science as a basic methodological discipline of a new type the form of which should always change as a result of the historical development of theoretical cognition and practice.

It is appropriate to emphasize that theoretical thinking and scientific cognition of every epoch develops in the shape of concrete contradiction of movement: it is always something *general*, it is a continuation, a sequence, a result of a certain general historic development of cognition and therefore it inevitably has certain features connecting it with the previous stages of the development of science but always represents something *specific* because it is a clear-cut product of the history of thinking (which is mediated by a number of *individual* intellectual, theoretical-cognitive and practical acts), cognition and human history in general. That is why materialist dialectics cannot fossilize into a certain form of generality but is must every now and then proceed to a concrete analysis of contemporary scientific cognition (and social practice). An excellent example in this sphere is provided for instance by Lenin's philosophical analysis of the crisis of physics which proved the untenability of ideas concerning the disappearence of matter caused by progress in physical learning from the level of atoms to electrons. Lenin thus formulated a significant methodological conclusion, fully corroborated by the development of physical learning, namely, that this advance of physical cognition would continue since the inexhaustibility of matter does not allow of any ideas concerning some elementary particles of matter which are indivisible any further. The development of scientific learning itself best demonstrated that Lenin's defence of *general* dialectical materialist principles is of fundamental theoretical and methodological importance for the subsequent *specific* levels of cognition in physics.

In the past few decades there has been a mounting interest in the specific trend of scientific studies which is sometimes described as a scientific movement; sometimes it is referred to as systems studies, theory of systems, *with systems analyses*, technologies, systems engineering, etc. being undertaken and developed. The main interest of scientists is now concentrated on complex systems, their analysis, designing, construction, functioning. Reality, its parts, spheres, levels, etc. *are approached* as systems (and subsystems). Consequently, we speak about the *systems approach.*

It is evident that in various branches of science, technology and human activity diverse forms and levels of the systems approach find a place. Systems research is being promoted in numerous directions. Mathematical, cybernetic, physical, biological, economic, linguistic, ecological and other systems are being investigated and projected. The systems approach is applied in the study of the behaviour of the organism just as in the solution of complex tasks in the field of technology, in the design and implementation of complicated technical systems, information systems, etc. The problems of automated systems, etc. are closely connected with these tasks.

Systems research projects also involve various walks of social life. It is particularly in these spheres in which the application of the systems approach is not only a purely scientific, theoretical and cognitive question but is closely linked with matters of social practice, with the issues of mankind's future. This finds its expression in endeavours to tackle the so-called global problems of the world today through the systems approach. Simultaneously, it is underscored that the systems approach is of paramount importance for the conscious comprehensive management of the development of society. The Marxist-Leninist theory proceeds from the real possibility and necessity of the *purposeful* comprehensive management of the development of society which is connected with the comprehensive transformation of the entire system of aspects of society's life, including the material, technical, economic, political, social, cultural, etc. A prerequisite for such a comprehensive transformation of the entire social system is formed by the purposeful management, by long-term influence by society as a whole which, however, can be based only on scientific forecasts and planning ahead into the future. Such an approach to the solution of the above-mentioned problems is impossible to apply in a society ruled by the partiality of interests caused by the principal elemental nature of capitalist circulation. Under such conditions efforts to enforce historical humanistic objectives encounter major barriers. This, in turn, leads to the adoption of limited technocratic approaches to the solution of problems of the development of the social system of the type of "theory of growth", "post-industrial society" or "social technologies" in the style of K. R. Popper.

These concepts continue to be based on the inability of grappling with the genuine human and social problems, of overcoming — along with the development of contemporary capitalism — the intertwined concept of "technological determinism" "which attempts to deduce immediate changes in the social structure, the political and ideological superstructure from the trends of technological and economic development while avoiding the cardinal question of the substance of social relationships, of the inevitability of fundamental changes in the forms of ownership, i.e. changes in the socio-economic system of bourgeois society".[1] It is only socialist society which is capable of consistently solving the problem of "creating *a comprehensive system of socio-economic indicators for all levels of the management of society*".[2]

With a view to the basic principles of Marxist methodology we should emphasize especially the paramount significance of the theoretical and methodological conclusions of the studies of social phenomena (as the supreme and most complex forms of the movement of matter) for the global solution of questions concerning the systems approach. Moreover, during the studies of social systems the basic contradiction distinguishing Marxist and non-Marxist approaches assume the form of reduction, the substance of which is a hypostasis, an overestimation, an exaggeration of a certain idealized systems orientation which may be elevated to the status of a certain world outlook or methodology. Thus, for example A. Rapoport believes that the general systems theory is not a theory at all according to the concept of this term in science but rather a world putlook or methodology[3] which is understood as a certain "orientation of exploration".[4]

A common feature of these concepts of the systems approach is the fact that they completely disregard the question of the relationship between objective and subjective dialectics (F. Engels), the question of the relationship between the objective reality and its reflection in consciousness (V. I. Lenin). Certain general logical-cognitive forms of cognition become independent of objects and processes of reality, of the specific logic of the specific object (K. Marx).

It is particularly in this sphere where we should find the methodological substance of errors by bourgeois authors when attempting to create comprehensive models of the development of society on the basis of the systems approach. (But as stated above, the entire issue also has deeper ideological connexions.)

In this sense a telling example was provided by the first report of the Rome Club, elaborated by D. Meadows and his collaborators, entitled The Limits to Growth[5] which was based on the so-called "systems dynamics" proposed by J. W. Forrester. The authors of this book attempted to construct a complex world-wide system characterized by an ever closer interconnexion among various aspects of human activity. As a result of this, human activity in one sector (for instance advancing industrialization) may lead to a catastrophe in another sector (pollution of the environment), etc. On the basis of a complex macro-system of movements within society as a whole on a world-wide scale the authors reached the conclusion that if the current trends of the growth of world population, industrialization, pollution, food production and depletion of natural resources continue unchecked, it is possible to anticipate — approximately in the following 100 years — that this planet will reach the "limits to growth". If a catastrophic collapse is to be prevented the authors recommended a transition to "zero growth". This was their formula on how to attain an equilibrium of the comprehensive world-wide social system.

Critiques both from the non-Marxist positions and especially those carried out by Marxist theoreticians demonstrated that the systems approach used in this case is absolutely untenable. At the same time, however, the critics admitted that the book did draw attention to very serious problems currently facing mankind. The critics also pointed to the fact that the application of mathematical models to such complex problems and such a complicated type of system as the development of mankind was premature or even impracticable. The use of complex computer technology created the impression of scientific foundations, "exactness" of employed methods, but in actual fact greater emphasis was laid on the systems dynamic machinery than on the real basis of the equations and formulas used. In other words, no attention was devoted to the theoretical and real points of departure of the "world model". The critics were justified in noting that the authors of the book operated within a somewhat unreal world system in which they distanced themselves from the specific features of social developments in the given historical epoch. This was reflected in their disregard for national frontiers, for differences in social systems, in their endeavours to create macro-systems without subsystems. As indicated by the title of the book itself, it is based on the premise of the bourgeois technocratic concept of continuous growth. This also constitutes the ideological as well as apologetic nature of this approach, which does not take into consideration the reality that scientific progress occurs in a real world containing two different social systems, the capitalist and socialist, which are characterized by different "trajectories" of development. In a certain sense it was already K. Marx who demonstrated that capitalist exploitation with all its consequences for the development of mankind has its own "limits of growth". A long time ago Marx showed that capitalism develops on a grand scale not only the material but also the subjective conditions of the labour process, but that it develops them as forces which dominate man and which are alien to him.[6]

The creation of "global models" of world development has registered a rapid upsurge in recent years. At the same time these models always have a systems character and their "construction" requires a comprehensive, systems approach. It is generally noted that the contemporary world development encounters no small obstacles including the problem of food caused by the rapid acceleration of the population growth, the problem of urbanization connected with the resettlement of populations into large city centres, the problem of man's relationship to the living environment linked up to ecological, energy-related and raw material difficulties, etc.

A great many of more recent global models elaborated by bourgeois authors have tried to incorporate social characteristics in the studied relationships, much thought is being given to the necessity of restructuring systems of objectives and orientation of social progress.[7] Even though these global models of world development of the third, fourth, fifth, etc. generation take into account the differences in the level of economic, political and cultural development of various spheres of present-day mankind, they never go beyond the confines set by the points of departure deeply rooted in the very nature of the capitalist social system. The authors of these concepts — under the guise of fighting the negative tendencies of scientific-technological and economic development which pose a common danger to all inhabitants of the earth — create a new variant of the idea of convergence of capitalism and socialism.

Global modelling of world development is closely linked with the question of prognostication of the development of systems of social relations. In this sphere, however, all the bourgeois global systems come up against the insurmountable obstacles (which they sometimes do not even realize to exist) given by the fact that in the capitalist world the aim of development and application of science and technology is first and foremost an economic growth which represents the development of capital, its expanded reproduction.

The Marxist-Leninist theory, quite understandably, devotes considerable attention to world-wide problems too. Addressing the 25th Congress of the Communist Party of the Soviet Union, its General Secretary, Leonid Brezhnev said: "Today there already exist considerably significant and topical global problems such as the problem of raw materials of energy, the eradication of the most dangerous and most widespread diseases, environmental protection, space exploration and the utilization of resources in the world's oceans. In tome, these problems will have an ever greater impact on the life of every nation, on the entire structure of international relations. Our country, just as the other socialist states, cannot stand idly by and not solve these problems which bear on the interests of all mankind."[8] Leonid Brezhnev and the other leading representatives of the socialist countries have on many previous occasions underlined that under the current conditions one of the prerequisites vital for the solution of these problems is the promotion of the process of easing international tension and the introduction of a lasting peace.

The process of real construction and continuous upsurge of advanced socialism both facilitates and necessitates the inclusion in the system of global relationship completely new, humane objectives and values corresponding to real socialism orienting and regulating the optimum development of man.[9]

Marxist methodology cannot ignore the fact that the systems nature is a characteristic quality, an aspect of objective reality and that as a result of this it is basically impossible to ignore the specific features of various forms and levels of objective systems, including social systems. Materialist dialectics, laws and principles of dialectical materialism require that the complex social systems should be studied from the viewpoint of their specific foundation, particularly while keeping in mind their concrete historical characteristic, i.e. class, socio-political, spiritual features, etc.

The very task of undertaking a dialectical-materialist analysis of any new scientific approach, and consequently the systems approach too, highlights the question of determination of this approach, of its specific character. The question may be formulated in the following manner too: what hase by the contemporary systems approach introduced that is new?

In specialized literature there exist numerous characteristic replies to this question, contribution to the clarification of the whole problem. For example, one non-Marxist representative of systems research, R. L. Ackoff who considers the term "system" to be the key concept of contemporary scientific studies, writes: "The tendency to study the system more as an entity than as a conglomerate of particles is parallel to the tendency of contemporary science not to study isolated phenomena in a narrowly limited context but rather in their open interactions so that ever broader spheres of nature could be examined. Thus we probably take part in the most concentrated efforts to accomplish a synthesis of scientific learning."[10]

Such a concept of the problem under investigation quite undoubtedly captyres some of the pronounced characteristic aspects of systems research projects. Simultaneously it is obvious that Ackoff presents in *reduced form* the characteristic process and in this connexion outlines the task which in a certain clear-cut form must have arisen already prior to scientific learning of the 19th century and was formulated by materialist dialectics. It is expecially the transition from the research into isolated phenomena to general connexions in the scientific learning of the 19th century that forms an inseparable part of the theoretical-methodological content of materialist dialectics. Particularly in his Dialectics of Nature, on the basis of the development of modern science, F. Engels substantiated the fact that science which in its early days first studies separate individual, abstractly isolated phenomena has reached the stage in which there emerge the dialectical connexions requiring the study of the overall connexions and mutual relationships ("open interactions" as R. L. Ackoff put it).

In this context it is approapriate to note some 100-year old ideas by Engels concerning the given problem. "Empirical natural science has accumulated such a tremendous mass of positive material for knowledge that the necessity of classifying it in each separate field of investigation systematically and in accordance with its inner inter-connection has become absolutely imperative. It is becoming equally imperative to bring the individual sphere of knowledge into the correct connection with one another... But it is precisely dialectics that constitutes the most important form of thinking for present-day natural science, for it alone offers the analogue for, and thereby occurring in nature, inter-connections in general, and transitions from one field of investigation to another."[11] F. Engels also explicitly formulated the idea of nature as *a system of connexions and processes.*[12]

Consequently, we view the question of the systems approach especially in this light; it is an issue of approach to mastering this system of *connexions and processes.*

But what is typical for such an approach which is designed to examine, clarify general laws and even to project and construct complex *systems of connexions and processes?* There exists a number of answers and opinions to this particular question and they cannot be dealt with here due to lack of time and space! But for the sake of the clarification of the question itself we would like to add two remarks.

1. Similarly to Ackoff, many other Marxist as well as non-Marxist oriented theoreticians have asserted that the systems approach is connected with efforts to attain a certain sythesis of scientific knowledge. To put this in other words, there is no doubt that the broadly based elaboration and application of systems approaches is linked with the current process of the integration of scientific learning. It is by no means indifferent to how this integration, this direction towards unification is understood. In this connexion, let us recall the fruitlessness of the neopositivist idea of a unified science which once appears as the idea of a unified language of science (and on the other hand, in Wittgenstein's later period it changed into its direct opposite, the idea of independent authentic language games), on another occasion as physicism, etc. It is expedient to note at least two outcomes of the development of the history of scientific knowledge. First and foremost it is the tendency towards a unified concept of the world which has always been connected in scientific cognition with its opposite, the differentiation of scientific learning. Integration unifies what has been differentiated. Engels demonstrated that the tendency towards the dialectical unity of lerning in natural sciences presupposed the accumulation of specific material, differentiated knowledge concerning individual aspects of the development of natural movement. In this context he underscored the origin of new marginal branches, viewed from the point of integrative cognition, (such as physical chemistry, biochemistry, etc.) which, represented steps forward both on the road to differentiation (origin, separation of the new scientific disciplines) but at the same time also in the process of integration (the uncovering of connexions existing among marginal scientific branches). In this connexion it is necessary to take into consideration the fact that in the history of science this contradictory movement of cognition (integration and differentiation) has been continuing and therefore various historical levels have to be distinguished, including all the detours and errors accompanying the process. Thus, for the period in which classic mechanics ruled supreme in modern sciences it may have

appeared that its general principles should have also prevailed in the scientific learning of other spheres of nature. Proponents of neo-positivist physicism, dazzled by the progress made by physics, have been following the road of the identical type of reductionism. Similarly, some theoreticians have been lately dazzled by cybernetics for example. As far as the present-day situation in the integration and differentation of scientific cognition is concerned, we hold the view that a conspicuous culmination, marked and initiated by the origin of Marxism, is currently underway. Marx and Engels established the foundations of integration and differentiation of scientific learning in society. The Marxist theory represents the intrinsic unity of the three component parts of the philosophy of dialectical and historical materialism, political economics and scientific socialism (communism), proceeding from the three major ideological sources, namely German classical philosophy, classical bourgeois political economics and utopian socialism.[13] This tripartite nucleus of the Marxist theory has laid down the basis for the unified theoretical as well as methodological conception of entire learning, in social sciences, history, jurisprudence, etc. Simultaneously — and this is very important indeed — it has established the real theoretical and methodological foundations for the integration of knowledge and cognition in social, natural and technical sciences. The unification of natural and historical (social) sciences was spelled out by Marx at a very early stage.[14] In later volumes Marx and Engels permanently reverted to these problems, especially in connexion with the studies of the social, natural and technical components of production as the decisive, determining sphere of the development of society, of human history. The process of interaction and overlapping of sciences is promoted and markedly combined during the process of the advancing scientific-technological revolution. There gradually arise new forms of mutual links among social, natural and technical sciences. "The process of the synthesis of scientific knowledge at present touch upon not only marginal disciplines but also on completely distant scientific branches which makes it possible to speak about the integration of scientific learning as one of the decisive factors of present-day developments."[15] The question involves not only the origin of new disciplines — ergonomy, technical aesthetics, engineering psychology etc., which penetrate into very distant branches, or the origin of such spheres as cybernetics, or the broad application of mathematics in various scientific disciplines, but first and foremost the fact that science now faces completely new branches marked by a considerably more versatile link-up to various other scientific disciplines. Figuratively speaking, it is as if the entire structure of the world so far, its present-day scientific image had disintegrated into an "unintelligible" unstructured mosaic and phenomena and aspects of objective reality and corresponding scientific knowledge are beginning to be "read", interpreted anew, creating new structures which make up a new image of the world. This is reflected especially in the formulation of global problems of mankind today. For example, present-day ecology which originated from biological ecology can hardly be characterized as a new scientific discipline in the traditional sense of the word and its nature can be best described as a new comprehensive type of science. The social, natural and technical aspects inherent in ecological problems manifest themselves as interacting, determing and penetrating one another.[16]

When analyzing systems approaches it is necessary to take into account the concrete, historical level of the development of scientific knowledge and interconnected historical tasks. At the same time when applying a certain systems approach it is impossible to ignore the enormous differences in the nature of studied objects, in their dynamics, complexity.

2. Particularly the programme concepts of the general systems theory tend to pose the problem (which is to a certain extent also contained in the concepts of the systems approach such as the general science theory and methodology) of whether it is at all possible to develop and formulate a methodology which would be both applicable and sufficiently fruitful for any system.[17] The problem can also be formulated as a question — whether it is useful to homogenize heterogenous systems or to put it more simply whether it is at all feasible to apply general characteristics of a formal or abstract system or as it is sometimes expressed, to "elaborate" certain general relationships on various arbitrary systems objects which posses very different substantial, concretely historical characteristics and which are located on completely different levels of forms of the movement of matter, etc. In this connexion we sometimes refer to two different systems approaches — the formal and dialectical ones. As regards the formal systems approach, there is the typical "definition by L. Bertalanffy according to which system is a complex of elements in interaction. Dialectical-systems approaches are characterized, on the contrary, by efforts to express as precisely as possible the specific features of the system as a whole as contrasted to its parts, consisting" — according to V. G. Afanasyev — "of the origin 'of new integrative properties which were not inherent as regards individual components making up the system'".[18] We consider it necessary to note that in this we encounter a specific form of the more general question which is typical for the development of contemporary science theories and methodology, notably the interrelationship between the formal idealizations the importance of which in current cognition is growing and the substantive expression of content in the reflection of the grasped objective reality. At the same time there can be no doubt what so ever that it is not only the content of our knowledge concerning objective reality that grows and is reinforced but also the means of logical, mathematical and other formal idealizations in scientific learning and their interpretation which are being developed. In the process of scientific cognition the formal and the substantive (form and content) are in the position of a dialectical unity: the expansion of the cognitive content in science is

accompanied by the enrichment of the formal means of cognition, their interaction, mutual enrichment and mutual penetration. Since this is not the place to continue discussing this question in greater detail, we assume the standpoint that in its development science is based especially on the existing content of learning and that a formalized theory cannot reflect the content of the scientific theory in question completely and exhaustively. Scientific knowledge, particularly at the present time, has been advancing thanks to the interaction of formal and substantive methods and means.[19] The gist of the whole matter is that in the current debate relating to the relationship between form and content, their role in the process of cognition reveals ever more clearly the specific integration of the two contradictory, differentiated trends in the development of scientific cognition which are characterized by the mutual penetration of the two contradictions. This specific type of integration of contemporary scientific cognition i.e. interaction, mutual influence, mutual penetration of the two contradictions, transition from the one into the other, is becoming an ever more topical problem of theoretical-cognitive and methodological research projects.

In our own opinion, for the investigation of the problem of the specific feature of contemporary systems research projects, the standpoint of V. P. Kuzmin is very fruitful and efficient to employ as he understands the systems nature as one of the major parametres, dimensions of the objective world in which scientific cognition takes place.[20] In his view, the current "explosion" of systems research projects involves an ever broader and deeper theoretical and practical clarification of this aspect of the phenomena of reality. Therefore, nowadays we use the term "system" to designate things which previously used to be described as entire, comprehensive, combined, organic, etc.

Undoubtedly, the rapid development of the systems approach is linked with the unprecedented upsurge of research projects pertaining to the behaviour of complex organized systems. This requirement appears both in the field of biological research projects and in the sphere of research into the "behaviour" and construction of complex structures in which growing complexity is being accompanied by the mounting importance of the problem of harmonization of the functioning of parts and the whole, management, self-regulation, self-control. A similar procedure is also necessary for the designing and planning of complicated processes in the social sphere. It should not be forgotten either that science itself, as compared with the past, has now become an immensely complex "organism". There arises the problem of the arrangement of the amassed wealth of knowledge, the further development of which lies in two mutually contradictory directions: the direction towards the differentiation and towards the integration of cognition (as implemented on this level). Various modes of efficient classification and utilization of the intricate sum-total of knowledge are currently being intensely sought.

The principles of materialist dialectics require that the study of phenomena of objective reality as well as the exploration of the forms of cognizing reality, categories, concepts and methodologies should be based on the unity of the logical and the historical. Cognition of objective reality itself, which is carried out along a specific road, constitutes a historical process which operates as a transition from non-existent knowledge to knowledge, from phenomenon to essence, from first-rate essence to secondrate essence, i.e. a transition from the lower grade to a higher one.

When Lenin, while studying Hegel's "Science of Logic" replied to the question "what constitutes dialectics?", he emphasized: "Every notion occurs in a certain relation, in a certain connection with all the others."[21] During the "materialist interpretation" of Hegel's dialectics, Lenin formulated this concept of the nature of cognition very pregnantly as follows: "Individual Being (an object, a phenomenon etc.) is (only) one side of the Idea (of truth). Truth requires still other sides of reality, which likewise appear only as independent and individual (*besonders für sich bestehende*). Only in their totality (*zusammen*), and in their relation (*Beziehung*) is truth realized."[22] In these words by Lenin we find significant guidelines for the solution of the problem under scrutiny. First and foremost, it is obvious that the problem of the specific character of the contemporary systems approach should be examined from the viewpoint of connexions between of the concept of the system and other concepts which play a key role in scientific learning and, simultaneously, that in the concept of the systems nature, the system reflects only one specific aspect of the phenomenon of objective reality and consequently assumes a special position in the sum-total of all the concepts. It is also necessary to keep in mind another notion of Lenin: "Man is confronted with a web of natural phenomena. Instinctive man, the savage, does not distinguish himself from nature. Conscious man does distinguish, categories are stages of distinguishing, i.e. of cognizing the world, focal points in the web, which assist in cognizing and mastering it."[23] In accordance with these (and other) requirements of materialist dialectics there appears the dual question: 1. what kind of position is assumed by the category system, the systems nature (the systems approach) in the totality, within the network of categories, and 2. on which particular level is the process of cognition situated at the moment?

What kind of position is thus assumed by the systems approach in the sum-total of all the other approaches towards phenomena of objective reality? According to the dialectical tradition, from Kant, Hegel, via Marx and Engels to Lenin, the position of a certain category, and consequently a methodological procedure based upon it, cannot be clarified without the systems approach which makes it possible to overcome the isolation of the methodological approach by incorporating it within a certain developing system (of categories, approaches). That is why one of the major aspects of materialist dialectics is the systems-categorial approach. A certain concept acquires a category status only if and when incorporated within a certain concrete system. It is clear that no final exploration of this particular problem can ensure that the demand for the concrete nature (which can be termed a certain ideal of cognition) would be met to the utmost. Lenin underlined this fact when formulating basic requirements of the dialectological determination of the subject of studies: "Firstly, if we are to have a true knowledge of an subject we must look at and examine all its facets, its connections and 'mediacies'. That is something we cannot ever hope to achieve completely, but the rule of comprehensiveness is a safeguard against mistakes and rigidity."[24] Individual research projects, however, are capable of testing the specific nature of the systems approach only when based on a certain "cluster", a group of categories.

It is expedient to note that particularly the philosophical principles and categories have for quite a long time been subjected to the systems approach. This was demonstrated in a way by G. W. F. Hegel in his philosophical system of absolute idealism. Let us quote at least this idea of Hegel: "Unsystematic philosophizing is not scientific. It may express an inner orientation or a private world view. If it is not justified as a moment of the whole and resting on unexamined premises or presuppositions, it possesses only a private inner certainty."[25]

Materialist dialectics links up to this systems concept of categories (methodological principles and procedures) and strives for the determination of their position, their "level" in the entire network of categories which has developed during the process of cognition. At this point we would like to mention what we consider as stimulating a work by V. S. Bibler, entitled "Concerning the System of Categories in Dialectical Logic", published in 1958.[26] We refer to this book not only because we want to demonstrate that in continuity to the works by Marx, Engels, and Lenin the systems approach in Marxist philosophy has been developed as if independently of the current boom of the systems approach, but also because we would like to underline some of the stimulating moments of this trend. First of all, the quoted book studies the content, relationship and role (function) of categories within the system of logical thinking and the process of cognition. The task of unfolding, deducing the system of categories is understood to be identical with the task of clarifying their coordinations. Man does not think in individual concepts but in their systems.[27] Moreover the process of abstraction, separation of individual categories is not a purely logical process; it relates to the history of science and technology, to man's practical activity. The system of categories cannot reflect the succession of the development of the world but the succession of the degreas of cognition and thus also the degrees of the objective transformation of the world by man.[28]

A very valuable application of the Marxist-Leninist tradition of the investigation of the systems of categories can be found in the works by V. P. Kuzmin[29] who documents his opinions both on the example of the historical development of the system approach in science and philosophy in general and on the example of the application of system principles of study in K. Marx in particular. According to Kuzmin the principle of the systems nature in cognition signifies the fact that phenomena of objective reality are studied as a law-abiding system whole and interaction of individual components making up the whole. The systems nature thus constitutes a special epistemological feature or a specific "dimension" of reality and despite the differences in the concept of "system" it is understood that the system represents a certain totality of elements which form a relative unity and entirety that possesses unifying qualities and laws. The basic problem, pertaining to the substantive as well as the methodological aspects, in the investigation of systems objectives is how to ascertain the determining features which are instrumental in the organization of elements. Thus this involves efforts to discover the specific foundation, relationships and connexions within the system, its specific quality and laws governing the structure, functioning and development of the given system.[30]

Kuzmin presents a characteristic of the systems approach on the basis of separation of three levels (types) of cognitive activity out of which each crystallizes around a certain principle that is based upon a certain idea concerning nature's typical objects. The simplest of the ideas considers as such a typical object of investigation an individual object, phenomenon, etc.; the more complex one: an entire system (macro or micro) and the most complicated one: reality as a totality of various objects. Corresponding to each level or type are specific viewpoints determining the mode of "viewing" the object. This means that first of all (1) in the 'focal point" of attention there is an individual object "by itself", then (2) "kind", "species" as systems of phenomena. The individual object or phenomenon appears here as a structural or functional component of the whole, as its element, i.e. the phenomenon is assigned to a certain system and subordinated to its laws. Finally (3), the subject of cognition is the complex reality as a numerous set of objects (systems) of various types and the phenomenon in question is studied synthetically and concretely historically as a product of the sumtotal of all operating conditions, external as well as internal.[31]

On the basis of these three different approaches to the study of reality, Kuzmin distinguishes three levels of theoretical reproduction of reality, as on all these three levels we find to a greater or lesser extent elements of the systems approach, but only the second level is predominantly a systems one.[32]

The first level — objective (immediate, simple one) of cognition of the individual, empirical, which appears as the immediately given, outside the causally genetic connexion, outside the system. The moment of the system entirety of cognition on this level is best noticeable in formal classification wherein each individual operates as an element of certain whole, kind, species.

The second level — system one (substantial)[33] of cognition investigated the phenomenon as an element of the system and ascertains the substantial basis of individual phenomena as a result of the study of the general conditions of the existence of individual phenomena and accordingly the emerging system character. The systems of phenomena are revealed by the history of the qualitative development of nature and society. The individual phenomena under scrutiny then operates as a moment of evolution, history. This gives rise to the following systems phenomena: pertaining to individual people — society and the laws of the development of socio-economic formations, pertaining to individual living organisms — biological species and the laws governing their evolution, etc. (for example in such pairs as the working class and the individual worker, society and personality, etc. — in all these cases the system e.g. the working class, determines the individual — the worker, representing its higher, system quality). The one and the same phenomenon is part and parcel of various systems of the real world. For instance, an individual man is a member of a certain class, a certain nation, a certain working team, etc. and as such possesses various specific qualities which, however, cannot be clarified on the basis of the man himself as a concrete individuality but only with the help of this system which determines his specific property. Unlike objective cognition in which the properties of the object (individual) depend on this object or individual alone, the systems level of cognition demonstrates that the object not only possesses numerous qualities but that also internally it is many-faceted as it contains various and diverse systems of objective reality.[34]

The third level — metasystem one (concrete, synthetic, "real") of cognition represents a transition towards theoretical reproduction of the entire object in the consciousness in many-sided unity of its real forms and interaction with other phenomena. This form of cognition operates with a totality of multi-system reality, with relationships of various order and dimension, it constitutes their synthesis. That is why it can clarify the origin, history of phenomena and facilitates prognostication on a scientific basis.

Looking at Kuzmin's concept of the qualitative specific nature of the systems approach, we can depict it in the following manner:

1st level	2nd level	3rd level
object by itself	systems qualities of objects with different levels	multi-system objects

At this point, we are obliged to emphasize that Kuzmin does not devote such attention to the third, multi-system level as to the second one. As far as the multi-system level is concerned, this, however, not only involve an analysis of complex, comprehensive systems on the basis of an analysis of subsystems, etc. This moment, to be true, plays an important role even though the main thing is that the systems approach which prevails on the second, systems level, should be supplemented by other approaches, especially by the genetic and historical ones without which the concretely thoughtful reproduction of concrete reality is impracticable in Marx's sense. Because, as Marx put it "the concrete is concrete since it is a totality of numerous specifications, hence a unity of the diverse".[35] In this sense the systems approach always remains somewhat one-sided, i.e. subjective and abstract. But this is a relative one-sideness unfolding in the direction towards the concrete[36], because the systems approach itself has been developing from the simplest outlines in an analysis of systems characteristics of reality's phenomena to very complex, comprehensive systems. This process has also given rise to and developed logical forms and methods of the systems approach, as well as a specialized language, while the means of the systems approach are differentiated on a very versatile scale.

In the concluding part we would like to return to the historical aspect of the problem under scrutiny. We believe that this aspect plays an irreplacable role in the study of scientific cognition and thinking. As F. Engels noted:
"An acquaintance with the historical course of evolution of human thought, with the views on the general inter-connections in the external world expressed at various times, is required by theoretical natural science for the additional reason that it furnishes a criterion of the theories propounded by this science itself. Here, however, lack of acquaintance with the history of philosophy is fairly frequently and glaringly displayed. Propositions which were advanced in philosophy centuries ago, which often enough have long been disposed of philosophically, are frequently put forward by theorizing natural scientists as brand-new wisdom and even become fashionable for a while."[37]

Our remarks on the historical development of the systems approach is inevitably only of a limited character; it is predominantly an attempt at demonstrating some of our previously mentioned comments.

We are also obliged to stipulate two preliminary reservations:

a) Proceeding from the concept of the systems level as outlined by V. P. Kuzmin, we should bear in mind even the prehistorical stage of the scientific systems concept of reality.

b) With a view to the above-mentioned classification of the formal and substantive in the development of cognition, we expressly state that we do not intend to deal with the special history of the formal aspect of the systems approach but especially its substantive side. Moreover, every effort is made not to mix up the substantive aspect with the dialectical concept. In our opinion such a mix-up of the contradiction between form and content and metaphysics (in the sense of a non-dialectical concept of reality) and dialectics could lead to confusion on the genuine problems related to theoretical cognition, methodology and world-outlook.

First of all, we will scrutinize the prehistoric, prescientific stage of development of the systems approach. This period is marked by the uncontrolled dialectical concept of the world as well as the concept of immediate experience of the world as a united, interconnected entirety and the gradual unravelling of interconnexions, links, influences of various spheres of objective reality.

Even a routine definition of the term "system" indicates (while taking into consideration all the historical and conceptual differences), that system is understood as a certain unity, connexion, a complex of phenomena, characterized by a certain order, an arrangement according to certain principles. In the original Greek sense of the word, system denotes — in actual fact — a whole made up of parts, something composed; the Stoics for example regarded it as a world order in the objective sense. Especially F. Engels singled out the importance of ancient Greek philosophy for the global concept of the woıld: "Among the Greeks — just because they were not yet advanced enough to dissect, analyze nature — nature is still viewed as a whole, in general. The universal connection of natural phenomena is not proved in regard to particulars; to the Greeks it is the result of direct contemplation. Herein lies the inadequacy of Greek philosophy, on account of which it had to yield later to other modes of outlook on the world. But herein also lies its supɹəiority over all its subsequent metaphysical opponents."[38]

Democritus' atomism represents an ingenious outline of the systems approach to reality. Aristotle's teachings also displayed elements of the systems concept of science.[39] It is necessary to note that ancient Greek philosophers linked up to older traditions and experience. Elements of the systems approach are to be found where ever, as a result of practical activity, theoreticians understood a certain order which unifies (integrates) the individual components into a certain whole and therefore sought principles and foundatio ns of these connexons. Thus, in ancient times people revealed "systems" connexions between the weather and harvest, between the movement of celestial bodies (the sun, earth) and the weather, etc. At that very inadequate level of the development of practice and learning these "systems" connexions were frequently explained with the help of supranatural principles and forces. In this context the system of myths and mythology which represent a "systems approach" *sui generis* would deserve a special analysis of its own. Unfortunately we must abandon such a project in this article. Let us only add that the mythology of advanced riverside civilizations (Egypt, Sumer, etc.) just as early ancient Greece constituted the primary ideological expression of the hierarchical "systems arrangement" developing within a class-divided society. Likewise the medieval anti-ecclesiastical and anti-feudal reformist and revolutionary movement, including the Hussite movement and its ideology, represented an attack against the hierarchically arranged system of feudal society.

The systems approach proper, which is much more developed, was promoted as late as in modern science. It was modern science that made an attempt to integrate learning on the basis of the unified natural system of mechanical bodies. It is modern science that developed, at the beginning implicitly and intuitively, the idea of the unified system of nature. Newton's laws of mechanical movement became a major presentation of the idea of "arrangement" of foundations, principles of systems connexions. But these principles of the natural system were based on the idea of unchangeability, the inner immobility of mechanical objects.

Metaphysical materialism was instrumental in the absolutization of the mechanicals concept of the world. Nature (including man) is conceived in this concept as a system of finished, unchanging particles, as their combination. For example Holbach in his book "The System of Nature" wrote that "nature in the broadest sense of the word represents a huge whole which is a result of the grouping of various combinations and various movements... "And further: "As a result of movement the whole establishes relations to its components and *vice versa*."[40]

This development results in the two following conclusions:

1. classic mechanics most probably constitutes the first form of a scientific system or science constructed upon the basis of an (undeveloped) systems approach to the mechanical level (form) of the movement of matter;

2. the metaphysical-materialist concept of the world represents the first attempt at a systems integration of scientific knowledge while the whole is studied as if composed of unchanging, given, mutually isolated components for which movement is, in actual fact, only an external characteristics. F. Engels said: "But what especially characterizes this period is the elaboration of a peculiar general outlook, the central point of which is the view of the absolute immutability of nature. In whatever way nature itself may have come into being, once present it remained as it was as long as it continued to exist."[41]

This corresponds to the attempt, for its time, at a unified classification, i.e. arrangement of plant and animal species as undertaken by Karl Linné in his book "The System of Nature".[42] In his interpretation, "the species of plants and animals had been established once for all when they came into existence; like continually produced like, and it was already a good deal for Linnaeus to have conceded that possibly here and there new species could have arisen by crossing".[43]

The entire period of the development of modern science has been marked by efforts to expand and promote the objective content of scientific cognition. But despite the numerous dialectical aspects which find their application and which occur during the process, this cognition has never arrived at an overall dialectical concept of reality. It is indisputable that we discern here certain fundamentals of the scientific systems approach which, however, can be termed neither as dialectical nor simply formal.

The task of elaborating the ideas of systems nature, the concept of system as an organically arranged, developing totality, was the task facing German classic philosophy which did so in an idealistic, speculative form. Kant had already totally overcome the notion that cognition is a mere mechanical sumtotal of ready pieces of knowledge and experience. He understood cognition as a certain arranged system with an inner order.[44] These efforts of classic German philosophy culminated in Hegel's teachings. We will leave aside the idealistic foundations of Hegel's philosophical system the principal materialist critique of which was undertaken by Marx and Engels. From our point of view the chief substantial aspect is that Hegel abolished any idea of elementary, indivisible, undeveloped, etc. given facts. For Hegel every given fact constitutes a certain whole, to wit a complex, a structured totality which is not merely a mechanical sum of its parts but rather an organic result of complicated developments. We demonstrated earlier that Hegel had applied his principles of the systems approach when expounding logical forms, concepts and categories as an inseparable aspect of his dialectical method.

Hegel did very much indeed for an understanding of the dialectical nature of the process of cognition. He understood this process as a complex dynamic, genetic system of moments and levels. Only in this way was it possible to realize genuine cognition. "*The systematic development of truth in scientific form can alone be the true shape in which truth existed.*"[45] Truth is the sum total which is attained as a result of thinking. Hegel criticized the concept according to which "*truth and falsehood* as commonly understood belong to those sharply defined ideas which claim a completely fixed nature of their own, one standing in solid isolation on this side, the other on that, without any community between them. Against that view it must be pointed out, that truth is not like stamped coin that is issued ready from the mint and so can be taken up and used".[46] Later on in the same book Hegel underlined this idea: "Dogmatism as a way of thinking, whether in ordinary knowledge or in the study of philosophy, is nothing else but the view that truth consists in a proposition, which is a fixed and final result, or again which is directly known."[47]

It is particularly the realization of the dialectical process like character of the system of human knowledge that represents the greatest accomplishment of Hegel's philosophy, and which was taken up by Marxism. Hegel also amply demostrated the abundance of dialectical connexions among levels and moments and moments of the overall system of learning and knowledge, he grasped the development of cognition as the unity of truth and faslhood, the immediate and the mediated, the abstract and the concrete, of the form and content, etc.

The system of knowledge was understood by Hegel as transient, changing.

This means the profound epoch-making significance of his dialectics which was an expression of the German philosophical revolution. The great contradiction of Hegel's philosophy, however, lay in the fact that on the one hand he understood truth as a dynamic system in its entirety, which he nevertheless presented as a closed one. These are the consequences of his absolute idealism. The entire system of truthful cognition can be presented neither in a single sentence, a set of sentences nor in one science. Hegel understood this very well, but he himself attempted to create such a system in his philosophy with the help of a speculative construction which was, however, only a finite one. Moreover, the development of natural sciences, not to mention social sciences, was — at Hegel's time — still in a very early stage. This can also explain the fact why Hegel highlighted the "principle of speculation" which — perhaps to the greatest possible extent — emphasized his idealism and which constituted the clear-cut form of Hegelian mystification of reality marked by "the objective nature of objects as a product of thinking and in a logicalized abstraction" concept "as a basic structure of all being".[48]

Thanks to the accumulated knowledge, the development of science in the 19th century made it possible to proceed to the study of natural and social phenomena from individual objects and elements of the totality of the world to their synthesis, while being aware of the general connexions and development. It was Marxist philosophy which presented a philosophical generalization of this progress in the development of cognition. In the middle of the past century there appear (independent of each other) two fundamental developmental theories characterized by the application of the systems approach to the study of a certain level (form) of movement of matter, namely Marx's socio-economic theory and Darwin's evolutionary theory. In both cases the theories represent a major contribution towards analysis of the development of the considerably large system of phenomena of objective reality. In the case of Marx, theory constitutes the foundation of the scientific concept of society. As regards Darwin's theory, its importance will be evident when compared with Linné's classification. It is proper to stress here that Marx did not elaborate a special methodology of the systems approach (just as he never wrote his intended study on the dialectical method), but he expounded it especially in his analyses of the capitalist manner of production. Marx regards society as an organic whole, which is not a linear level but a totality that develops according to the laws laid down by the given socio-economic formation. Marx revealed the systems character of the socio-economic formation in general and the capitalist one in particular. This means that when studying social phenomena, Marx proceeded from the basic properties of the formation in question which belong to it as a certain concrete organic system. He placed individual social phenomena in a relationship to this system and clarified them according to its laws. Marx did admit certain social traits within the development of society but he was chiefly concerned with uncovering the specific law of development of one particular society, the capitalist one, the capitalist system. He disapproved of the general structure belonging to formations without exception because principally different structural laws correspond to various social systems in various stages of historical development.

Marx described his method of approach to the study of the capitalist manner of production (system) as sufficiently distinctive. Let us recall here only some ideas. First and foremost, he demonstrated that "all epochs of production possess some common features, common specifications. Production in general is an abstraction but a sensible abstraction insofar as it genuinely underscores what is common, fixes it and prevents repetition". But "determinations which apply to production in general must be excluded owing to the fact that because of unity — which proceeds from the fact that subject, mankind, and object, nature are identical — the substantial difference should not be forgotten".[49] Marx then concluded his methodological remarks as follows: "There are specifications, common to all the levels of production which are fixed by thinking as generally valid, but the so-called general conditions of the entire production are nothing else but these abstract moments through which we can depict no genuine historical level of production."[50]

We hold the view that analogically it is possible to refer to the fruitfulness of attempts at separation of the system and structural principles in a certain general scientific level which leads inevitably to the development of elements of the systems approach. It should be evident too that this cannot replace other approaches to the synthesis of scientific knowledge and also it is necessary to underline that the laws governing individual systems of phenomena are always specific for the given developmental level of reality, for the given form of matter, etc.

V. P. Kuzmin presented a profound analysis of Marx's systems approach. Therefore we would like to single out one particular aspect of his investigation of the problem under review, which however can be interpreted here only in a very simplified manner.

Kuzmin sees as the main merit of Marx's discovery of the systems nature of social reality his disclosure of the dual qualitative determination of social phenomena, be it the complex discovery of the dual character of labour followed up by the discovery of the so called dual character of value, its the bone discovery of the so-called dual determination of natural and social qualities according to their material-structural and socio-functional specification.[51] (Let us only add that this dual nature of determination, natural and social, one, is contained in Marx's concept of production).

The gist of Marx's discovery lies in that all products of human labour retain their natural qualities but acquire social properties as well which determine their social function and their development. Every product of human labour (the simplest manufactured object as well as the most sophisticated technical equipment, just as so-called cultural plants or domestic animals, etc.) contains its material structure, being subordinated to the laws governing the specific form of matter while this specific natural quality of products of human labour is forced in to the background (it is dialectically negated) by their main social quality. Primary natural properties of human products are in fact means of existence of their social qualities. These social qualities exist independently of man, while the social qualities were created by human labour and their social life is determined by their function within the system of society.

Marx distinguished between two degrees of social qualities. Material products of human labour possess certain material qualities which are instrumental in making them utility objects, in creating utility value (they are the results of concrete human labour). Social qualities of a second order cannot be deduced directly from individual objects but they express integrating qualities of the social whole, of the social systems. These qualities are impossible to individualize in terms of an empirically concrete object, but they reflect a certain general property of a number of phenomena. These are systems qualitative determinations. An incorrect understanding of this fact may result in material fetishism which endeavours to seek in individual phenomena that which forms the systems reality. In this connexion Marx referred to the fact that as regards value not an atom of matter enters into its composition.[52] How could Marx find in a commodity such a complex and "variable" quality, as value which does and does not relate to the commodity, appears as a basic feature of every kind of the commodity and simultaneously is not included in it materially? Marx solved the problem on the basis of the discovery of social systems quality which determines, provides the possibility of measuring the mutual relationship of individual utility values, individual types of commodities. The value of a commodity is determined by the fact that utility value, its material form contains and materializes abstract human labour. According to this principle a whole range of social connexions which are situated outside the sphere of control of operating people and which possess the nature of natural relationships has been developed.

The merit of Marx's analysis lay in that value was seen as a social quality of commodities, as a system quality of a sum of commodities. This gave rise to the tripartite qualitative nature of one single thing, a commodity, namely its natural quality, social quality of the first order and social quality of the second order. The main systems quality is then formed by value as the most supreme systems quality.

We have already mentioned that independently of Marx's clarification of the nature of the social system, Charles Darwin's evolutionary theory shed light on the genetic, developmental character of animate nature in the shape of *a system of connexions* and transitions.

But the 19th century yielded yet another major discovery in natural sciences: the discovery of the periodic law of chemical elements and the formation of the *periodic system of chemical elements* by D. I. Mendeleyev. An analysis of this system will enable us to demonstrate some other aspects of the concept of the systems approach under scrutiny.[53] Our purpose here is not to clarify the importance of the periodic law and the periodic table of chemical elements. There can be no doubt whatsoever that the law became the basis in its field, a foundation for further discoveries of new elements, for the exploration of the inner structure of atoms' outer shells and latter on of their nuclei and their transformations, synthesis of trans-uranium elements, etc. Indisputably, Mendeleyev was no conscious dialectician. "In his works, dialectics appeared only in an uncontrolled manner because it was related to the studied object as well as to the process of the study of this subject."[54] But we are mainly concerned with the fact that consciously empliying the scientific means available to chemistry of his time (utilizing the theoretical and empirical methods of analysis and synthesis), Mendeleyev tackled the problem of the relationship between the element and the system, the systems connexions. The character of the element in question determines its "position" within the entire system and the system as its genetic connexion of transitions.

Attempts at classification of chemical elements were undertaken prior to the discovery by D. I. Mendeleyev. Let us name at least J. W. Dohereiner, Bégnyer de Chancourtoise and J. A. R. Neulands. They and many others proceeded from similar or identical qualities of individual chemical elements. First steps in the research of chemical elements revealed especially differences among them. On this basis it was feasible to describe only some of their external characteristics, not their internal connexions. The substance of Mendeleyev's discovery is the disclosure of the inner unity of chemical elements.

For the sake of brevity we will list here only some, for us, substantial circumstances, key items in Kedrov's brief summary of Mendeleyev's discovery. "Mendeleyev's chief idea was the necessarity to place into mutual relationships and connexions not only identical but also *non-identical* (different) chemical elements, including diametrically opposed elements such as alkaline metals (chemically most active metals) and halogens (equally active non-metals). The first and the second groups form two extremes in the set of all chemical elements and their affinity (unity) in particular, which had previously not been admitted; they expressed the unity and mutual penetration of contradictions which determined the entire course of the discovery."[55] Mendeleyev based his analysis on the observation of two elements Cl and K and their affinity established by their *atomic weights:* $Cl = 35{,}5$, $K = 39$. The comparison of these two chemically extreme elements is described by Kedrov as the "cell" of the entire discovery because this contained the fundamental contradiction of the very object of research (chemical elements). The contradictory relationship between the elements belonging to the two polar groups led Mendeleyev to the discovery of the analogical relation existing among whole groups of elements. By comparing groups of alkaline metals and the group of halogens — according to the size of atomic weight of the elements of these groups — he arrived at the expansion of the original "cell".

The two lines then looked as follows:

$Li = 7$	$Na = 28$	$K = 39$	$Re = 85{,}4$	$Cr = 133$
	$F = 19$	$Cl = 35{,}5$	$Br = 80$	$J = 127$

which was justifiably considered by Kedrov as "outlining the core of the future system".[56] We can easily state today that the comparison and determination of these two subsystems played the key role in the *genesis* of the entire system.

It is also vital to underline the fact that for Mendeleyev the comparison of atomic weights of elements constituted a point of departure because at the outset atomic weights were the only general feature according to which it was feasible to compare all different chemical elements. "But according to how the compared groups gradually resulted in the formation of firm foundations of the future system of elements, taking into account all their incompleteness, there starts to emerge a new feature which characterized — owing to its substance — a certain chemical element not as an individual itself but as *the individual within the general*, i.e. as an element which found its *place* in the overall system of all elements."[57] According to Kedrov, particularly the formation of this new feature of the "position" of the element within the system made it possible to complete the construction of the system. It does not express only a summary of properties of a certain particular element but a) the relationships of a given element to the "neighbouring" elements within the system and b) through them to the entire system. "This means that the feature of the 'position' captures both '*the nature of a given element* (a sum of its properties) and also its *interconnexions with other elements* (a sum of its relationships with them)'."[58]

It was especially the understanding of the relationship between "element" and "system" which enabled Mendeleyev to solve a number of problems stemming from the then state of theoretical and empirical studies of chemical elements. This involved, for example, the theoretical investigation of the rate of errors in experimentally determined atomic weights of some elements, the discovery that some "unoccupied places" in the system must be occupied by elements not yet known whose atomic weights Mendeleyev established on the basis of connexions of the system (gallium, rhodium, germanium), the correct determination of the order ("positions" within the system) even of those elements whose atomic weights were lower than the atomic weights of elements situated in the system in front of them (so-called anomalies: tellur—iodium, cobalt—nickel), etc.

In order to underline the strength of Mendeleyev's uncontrolled dialectical concept of the inner connexions within the system of chemical elements, we quote here the methodologically very stimulating idea of A. V. Rakovsky, which dates back to the year 1927. He wrote then: "Let us image ourselves living in 1871 with our own psychology, with our own manner of respecting experimental data; we are approached by a chemist who proclaims that he has discovered a periodic law and that according to this law he is building a natural system of elements but that in order to do this out of 64 elements (the number of elements known at that time — note by the author), 28 have to be subjected to minor or major violence or radical changes of their atomic weights. I would be hardly wrong if I say that most of us would assume a negative attitude to such a discovery and would regard this system as unnatural. It is indisputable that Mendeleyev did not discover his system by blindly following facts but on the basis of a critical analysis of them."[59]

It should be added here that in actual fact in the substance of Mendeleyev's discovery we can discern a remarkable affinity (unintentional, uncontrolled) with Marx's method of transition from the abstractly general to the concretely general which forms the foundation of the discovery of the inner connexions, the concrete basis of the systems connexions linking phenomena of objective reality.

Speaking purely abstractly, even prior to Mendeleyev's discovery it was known that atomic weight is a general characteristic of all chemical elements. But this characteristic appeared merely as a "mute" generality. It was Mendeleyev to whom this characteristic "spoke" for the first time as a concrete generality. What — in brief — does this aspect involve? Mendeleyev understood a chemical element as a concrete specific sum total of organically interlinked properties. Having carried out a comparison of the entire set of properties of various elements, he arrived at the conclusion that the atomic weight, or more exactly the "position" of an element within the system is the determining characteristic of every element. At the same time it is necessary to realize that this involves on the one hand the relation between the atomic weight as a quantitative specification of function and its qualitative chemical properties but, on the other hand, behind this relation there appears a new characteristic, a determining systems quality, the "position" of the element within the system which cannot be deduced from any individual element and which — analogically to the Marxian case of the analysis of the value of commodity — does not contain an "atom of matter" but rather expresses the systems connexion among the masses of individual phenomena. As a matter of fact, Mendeleyev discovered the dual qualitative nature of chemical elements, the fact that every element — in addition to its empirical qualitative specification — possesses another, more profound systems quality.

Let us make yet another attempt to express this idea a different way. Atomic weight is a characteristic common to all elements. But as such it is only a secondary, abstract generality. Simultaneously, it is a quality specific to every element, a concretely quantitatively determined property. It is a property which abstractly connects (incorporates) all elements but at the same time it concretely divides, distinguishes every individual element from all the others. Hence, it is such a concretely general property, while the general — as Marx expressed it — "is itself divided into many facets and opens out into various specifications". On this basis Mendeleyev created a genetic system of chemical elements, arranged from the simplest, the most elementary one to the most complex, wherein every element assumes its own "regular position" and appears as a degree, a moment of the entire system.[60]

In our remarks we focused attention especially on efforts to make a contribution towards clarifying certain matters connected with an approach to reality which understands and examines it as *a system of connexions, processes and transitions, revealing genetic and developmental connexions* of phenomena of objective reality by ascertaining the determining quality of any given system (systems quality). At the same time we do not ignore the necessity and fruitfulness of other, for instance, formalized systems approaches in contemporary scientific cognition.

REFERENCES

1 "Sotsiologicheskie issledovaniya: rezultaty, problemy i zadachi" (Sociological Research, Results, Problems, Tasks), *Kommunist* (Moscow), September 1980, p. 91.

2 Ibid., p. 88.

3 See A. Rapoport, "Matematicheskiye aspekty abstraktnogo analiza sistem", (Mathematical Aspects of the Abstract Analysis of System), in: *Issledovaniya po obshchei teorii sistem* (sbornik), Moscow 1969, p. 83.

4 See A. Rapoport, "Razlichnye podkhody k obshchei teorii sistem. Sistemnye issledovaniya" (Various Approaches to the General System Theory. System Studies), in: *Jezhegodnik*, 1969, p. 55 and 61.

5 D. Meadows and a team of authors, *The Limits to Growth*, New York 1972.

6 See K. Marx, *Theories of Surplus-Value*, Moscow 1973, p. 379.

7 See e.g. E. Laszlo, *A Strategy for the Future: The System Approach to World Order*, New York 1974; W. Leontieff, *The Future of the World Economy*, New York 1976.

8 *The 25th Congress of the Communist Party of the Soviet Union*, Prague 1977, p. 52 (In Czech).

9 Critical Marxist-Leninist assessments of the concepts of global models can be found in the following works: V. Zagladin—I. Frolov, "Today's Global Problems and the Communists", *World Marxist Review*, 1978, No. 3, pp. 64)74, and D. M. Gvishiani, "Global Modelling: A Comprehensive Analysis of World Development", *World Marxist Review*, 1978, No. 8, pp. 59—63. Also see D. M. Gvishiani, "Metodologicheskiye problemy modelirovaniya globalnogo razvitiya" (Methodological Problems

of Modelling Global Development), in: *Voprossi filosofii*, 1978, No. 2, pp. 3–13.

10 R. L. Ackoff, Games, Decisions and Organizations, in: *General Systems*, vol. IV, 1959, p. 145.

11 F. Engels, *Dialectics of Nature*, Moscow 1972, p. 42.

12 "Today the whole of nature lies spread out before us as a system of inter-connections and processes that, at least in its main features, has been explained and understood" (Ibid., p. 198).

13 Comp.: V. I. Lenin, "Three Sources and Three Component Parts of Marxism", *Collected Works*, vol. 19, Moscow 1977, pp. 23–27.

14 "History itself is a *real* part of *natural history*—of nature's coming to be man. Natural science will in time subsume under itself the science of man, just as the science of man will subsume under itself natural science: there will be *one* science." (K. Marx, *Economic and Philosophic Manuscripts of 1844*, Moscow (1967, p. 103).

15 P. N. Fedoseyev, "Filosofiya i integratsiya znaniya" (Philosophy and Integration of cognition), in: *Voprosi filosofii*, 1978, No 7, p. 20.

16 See R. Kolářský, "Dialektika člověka a prostředí" (The Dialectics of Man and Environment), Filosofický časopis No 4/1981, pp. 525–534.

17 During discussions concerning the possibility of utilizing the systems approach in the research of the system of science, K. Berka pointed to this problem in his article "Theoretisch-methodologische Bemerkungen zur Systemauffassung in der Wissenschaft". (See *Teorie a metoda* (Theory and Method), Prague, 1973, No 2, pp. 83–97).

18 P. Ernst, "Aktuálnost systémového výzkumu a zdokonalování plánovitého řízení VTR" (The Topicality of System Resesarch and Improvement of the Planned Management of Scientific Technological Progress), in: *Teorie rozvoje vědy* (Theory of Science Development, Prague, 1980, No 3, p. 54).

19 See V. I. Kurayev, "Leninskoe ponimanie poznaniya i problem rosta znaniya" (The Leninist Concept of Cognition and the Problem of the Growth of Learning), in: *Voprossi filosofii*, 1980, No 4, pp. 100–111.

20 See V. P. Kuzmin, *Printsipy sistemnosti v teorii i metodologii K. Marksa* (The Principles of Systems in the Theory and Methodology of Karl Marx), Moscow 1976.

21 V. I. Lenin, *Collected Works*, vel. 38, Moscow 1961, p. 197.

22 Ibid., pp. 195–196.

23 V. I. Lenin, *Collected Works*, vel. 38, Moscow 1961, p. 93.

24 V. I. Lenin, *Collected Works*, vel. 32, Moscow 1977, p. 94.

25 G. W. F. Hegel, *Encyclopaedia of Philosophy*, New York 1959, p. 72.

25 V. S. Bibler, *O sisteme kategoriy dialekticheskoi logiki* (Concerning the System of Categories in Dialectical Logic), Stalinabad, 1958.

27 Ibid., p. 6–11.

28 Ibid., pp. 12, 13.

29 See V. P. Kuzmin, *Printsipy sistemnosti v teorii i metodologii K. Marksa* (The Principles of Systems in the Theory and Methodology of Karl Marx), Moscow 1976.

30 Ibid., pp. 10–11.

31 See V. P. Kuzmin, *Printsipy sistemnosti v teorii i metodologii K. Marksa*, Moscow 1976, pp. 30–31, 36–37.

32 Ibid., pp. 37–66.

33 But Kuzmin understands substance as a specific material foundation of a group of phenomena.

34 Particularly this aspect was very pointedly underlined by K. Marx when he proved that human qualities (and their changes) are dependent on those material circumstances whose parts the people in question are.

[35] K. Marx, *Grundrisse der Kritik der politischen Ökonomie*, Berlin 1953, p. 21 („Das Konkrete ist konkret weil es die Zusammenfassung vieler Bestimmungen ist, also Einheit des Mannigfaltigen".)

[36] "Logical concepts are subjective so long as they remain 'abstract' in their abstract form, but at the same time they express also the things-in-themselves. Nature in both concrete and abstract, both phenomenon and essence, both moment and relation. Human concepts are subjective in their abstractness, separateness, but objective as a whole, in the process, in the sum-total, in the tendency, in the source." (V. I. Lenin, *Collected Works*, vol. 38, Moscow 1961, p. 208).

[37] F. Engels, *Dialectics of Nature*, Moscow 1972, p. 43.

[38] F. Engels, *Dialectics of Nature*, Moscow 1972, p. 46.

[39] See K. Berka, "Theoretisch-methodologische Bemerkungen zur Systemauffassung in der Wissenschaft", in: *Teorie a metoda* (Theory and Method), 1973, No 2, p. 83 ff.

[40] P. H. Holbach, *System der Natur*, Berlin 1960, pp. 17, 46.

[41] F. Engels, *Dialectics of Nature*, Moscow 1972, p. 24.

[42] It is quite interesting how the 1900 "Otto Encyclopaedic Dictionary" (Volume XVI, p. 53) characterized this book. The encyclopaedia said that Linné "classified organisms according to broader and narrower spheres of characteristics, expressing their broader and narrower affinity, into kinds, species, orders and classes and eventually arranged these categories—according to a precisely observed method—in an upward manner into a classified, systematic whole".

[43] F. Engels, *Dialectics of Nature*, Moscow 1972, p. 24.

[44] "A theory, if it is to be a system, i.e. a totality of learning arranged according to principles, is called science... These principles can constitute either an empirical or rational connexion of knowledge into a whole." (I. Kant, "Metaphysische Anfangsgrunde der Naturwissenschaft", in: *Gesammelte Schriften*, volume 4, Berlin 1911, p. 467).

[45] G. W. F. Hegel, *The Phenomenology of Mind*, London 1910, p. 5.

[46] Ibid., p. 46.

[47] Ibid., p. 37.

[48] J. Zelený, *Verstand und Vernunft in Hegels „Wissenschaft der Logik" und in der materialistischen Dialektik*, Prague 1980, pp. 12—13.

[49] K. Marx, *Grundrisse der Kritik der politischen Ökonomie*, Berlin 1953, p. 7.

[50] Ibid., p. 10.

[51] V. P. Kuzmin, *Printsipy sistemnosti v teorii i metodologii K. Marksa* (The Principles of the Systems Nature in the Theory and Methodology of Karl Marx), Moscow 1976, pp. 73—74.

[52] See K. Marx *Capital*, volume I, Moscow 1971 p. 54.

[53] B. M. Kedrov a distinguished Soviet theoretician has been engaged for years in a dialectical materialist analysis of Mendeleyev's discovery. His investigations represent an outstanding demonstration of the dialectical nature and substance of cognition in natural sciences which is a reflection of the dialectical connexions of the material world. In this study, apart from our own analyses we draw predominatly on Kedrov's study A "Dialectical Analysis of a Great Scientific Discovery". To mark the centenary of the discovery of the periodic law of chemical elements, the study was published in the book by B. M. Kedrov, *Filozofia vedy* (Philosophy of Science), Bratislava 1980, pp. 288—308; B. M. Kedrov, "Dialekticheskiy analiz velikogo otkrytiya (A Dialectical Analysis of a Great Discovery), in: *Voprosi filosofii*, 1969, No 3.—As far as B. M. Kedrov's other major works devoted to this problem are concerned, we name at least the following monographs: *Den odnogo velikogo otkrytiya*

(The Day of One Great Discovery), Moscow 1958, 560 p.; *Filosofskiy analiz pervych trudov D. I. Mendelyeeva o periodicheskom zakone* (A Philosophical Analysis of D. I. Mendeleyev's Early Works Concerning the Periodic Law), (1869–1871), Moscow 1959, 294 p.; *Mikroanatomiya velikogo otkrytiya.* K *100-letiyu zakona Mendeleyeva* (A Micro-anatomy of a Great Discovery. In Honour of the Centenary of the Mendeleyev Law), Moscow 1970, 243 p.

54 B. M. Kedrov, *Filozofia vedy*, Bratislava 1980, p. 290.

55 Ibid., pp. 290–291.

56 See ibid., p. 297.

57 Ibid., p. 300.

58 Ibid., p. 302.

59 According to the miscellany *Periodichiskiy zakon D. I. Mendeleyeva i ego filosoficheskoe znachenie* (Mendeleyev's Periodic Law and its Philosophical Meaning), Moscow 1947, p. 42.

60 As concerns this aspect of Mendeleyev's periodic system of chemical elements in greater detail, cf. Z. Javůrek, *Dialektika obecného a zvláštního* (Dialectics of the General and the Particular), Prague 1962, pp. 59–64. (This study did not proceed specifically from the systems approach but from the problems of the dialectical relation of the general, special and particular.)

10. The Role of General Scientific Concepts in the Integration Process in Science

JIŘINA STACHOVÁ

The formation and transformation of concepts represents one of the inner mechanisms of scientific cognition. Concepts, like key moments in the network of reality's reflection, originate as instruments of the subject on its way to penetrating the substantial connections of the surrounding world. Concepts make up a changing network developed on numerous levels of abstracting and generalizing the activity carried out by man as a social being. The formation and evolution of concepts reflects the complexity and contradictory nature of comprehending what is general and substantial. It demonstrates that the path of cognition, of penetrating into the substantial and general connections of objective reality, is far from being straightforward. Similarly as the practical transformation of the world for the benefit of man, for the benefit of society is not simple and straightforward. This bears out Lenin's words "Practice is higher than (theoretical) knowledge, for it has not only the dignity of universality, but also of immediate actuality".[1]

The system of concepts in the process of scientific cognition is being formed and transformed in the conscious, inevitable endeavours of mankind to capture the character and regularities of individual spheres of reality and their interrelations as well. This process rationalizes practice itself, accompanying it by the contradictory process of social development in which not only practice is being rationalized but cognition too is being made practical. This brings them together and their own specific creativity is simultaneously multiplied. Scientific concepts in this process play the important role of cornerstones in which the individual levels of capturing reality are grasped, fixed and unified. They comprise the very contradictoriness of the social process of cognition, including its links with practice, while separated from this real historical process they offer themselves for misuse, absolutization, mystification.

The gnoseological moment and sense of the conceptual reflection of reality is promoted through the growing activity and creativity of the social subject. The problem of the active, creative role of the cognizing subject represents one of the major problems presented by modern science. This does not involve an abstract "individual" subject but a concrete socially determined cognizing subject, comprising in a historically feasible manner the "practical nature" of reason and the "rationality" of practice in the purposeful direction towards the perfection of them separately, as well as in their unity. The idea that science develops from itself as an evolving abstract reason will not be accepted by modern science in any modification. The conceptual network of science is established, differentiated and stratified in order to capture individual images of reality in an actual and creative activity of the recognizing subject. "Man's consciousness not only reflects the objective world, but creates it."[2] In these words Lenin expressed the very nature of the dialectics of human creative activity in the unity of its objective and subjective aspects.

Scientific concepts and categories themselves contain the contradictoriness of this activity. They constitute a mediating moment of fundamental developmental trends of science as well as their product. They are the direct agents in the process of differentiation and integration, the carriers of analogy, similarity and differences, they have their relevance to scientific revolutions as well. In this paper we focus attention on the role of scientific concepts in the integrative process in sicence. The integration of cognition represents one of the regular processes in the development of science. In the evolution of science, integration tendencies have assumed various forms and, naturally, are affected in their manifestations by the given historic possibilities. Very often the integration of science is considered only as a product of the development of modern science, particularly its contemporary stage. The integration of knowledge, however, is an equally old process as its opposite, namely the differentiation of knowledge. Both tendencies proceed — from the nature of the grasped objective reality mediated through the character and level of social practice. Above all, they result from the dialectics of continuity and discontinuity of reality. We do not want to deny that in the history of science the intensity and quality of both tendencies were different. The same applies to contemporary sicence. It is the process of integration which facilitates, if and when its mechanism is thoroughly analyzed and utilized in the management of the development of science in an adequate manner, a substantial speeding up to the development of science and a major impulse of its orientation. This would place knowledge as a socio-historical process, as an instrument for the solution of the contradictions of society and nature in its "existential" form onto a new, qualitatively higher level which would, in turn, considerably change the contradictions in question. This process is tied up with a number of conditions and prerequisites, notably of a social character.

The differentiation and integration of science takes place not only horizontally, i.e. in the form of the branching off and merging of disciplines, but also vertically in the sense of various levels of abstraction of reality and consequently in the sense of different mediations in respect to social practice.

The dialectic of differentiation and integration within the two abovementioned levels is projected in the conceptual sphere of science particuarly as the dialectic of the general and the special. A special concept of a specific discipline at the same time carries a very general categorial content given by a complex link with the general theoretical basis of the scientific branch in question. On the contrary concepts of the disciplines that are results of the integrative process inevitably contain special moments due to the specific charcter of the processes and the laws studied by these sciences. For instance, the concept "spin" is a special concept of nuclear physics. As far as its content is concerned, however, it is quite inevitably linked up to the current theory of elementary particles. Its content forms the unity of the special, reflecting something which is a property only of certain phenomena of the microworld, and the general which enters its content from physical theories. At the same time, these physical theories are necessarily linked to the general knowledge of the physical nature of the world. The special and the general tend to enrich one another while a mutual contradiction, whose solution constitutes one of the stimuli for overcoming the attained level of knowledge and the origin of new theories, is growing. Analogically, on the other hand such concepts as "system", "structure", etc., even though reflecting the very general aspects of reality, are simultaneously an expression of a certain specific approach to this reality. They lay down the frame for adding details to the picture of objective reality.

Cognition proceeds from the apprehension of the identical, indistinguishable to the different, distinguished. It leads to the cognition of further specific properties which complete the qualitative definiteness and consequently the distinction of originally undistinguished phenomena. On the other hand, it proceeds from the distinguished phenomena to the apprehension of similar, even identical, ones. These two processes determine one another and multiply their effectiveness. They are present also in the process of the formation of concepts in the unity of continuity and discontinuity of changes in the contents of the existing concepts. The clashes of the two aspects find their expression in substantial changes in the contents of concepts and lead to the formation of new scientific concepts of the new theory which deepens our cognition of the reality in question.

To understand the nature of the process of concepts formation in science is of considerable methodological-epistemological significance. Each and every scientific concept combines in itself two moments: the fixative and predicative one. In respect to the first one scientific concepts fix within the structure of interconnexions in the given theory the attained level of knowledge. In respect to the second one scientific concepts include, at the same time, new knowledge, which enrich its contents. As a consequence of such concretization of knowledge concepts get into a clash with the theory in question, and become, thus, one of the stimuli of the gradual modifications and eventual dialectical negation of the theory itself. This process is rather complex, it has many different degrees, with a different dynamics rate of acceleration and numerous random events. The dialectic of theory and concepts through which it is expressed, the mutual connexions and contradictions, represent one of the motives for the progress of scientific cognition. The relationship between theory and "its" concepts is not conceived from the epistemological point of view as a relationship between the whole and its parts or between the more complex and the simple. In theory each concept appears in only one particular concrete link while its epistemological value can be implemented in many other "theoretical" links. It is, therefore, richer than as utilized by any concrete theory. On the contrary, theory multiplies the epistemological value of concepts by placing them in a certain conceptual structure, thus "revitalizing" or co-forming them or rather completing them in the direction of the given knowledge. Theories constitute a "nourishing milieu" in which the cognizing function of concepts is realized. In the relationship between theory and concepts on different levels the subject-object relation is implemented. This relationship has its own complex background from which it proceeds and within which it is formed. The basic characteristics of the process of cognition, cognizing activity as a subjective reflection of objective reality assume in scientific cognition their specific form as a result of the proliferation of conscious subjective creative activity. This activity represents a regular, inevitable prerequisite for the solution of the fundamental relationship between society and nature in its objective shape. The multi-level nature of the epistemological relation between subject and reality finds its expression also in the multi-level character of the creative moment in scientific cognition as a reflection of reality. The determination of the object of cognition is the primary basic degree of the complicated process of the epistemological and not only epistemological separation of man from reality, from the reality existing independently of man's consciousness. In this stage of reflection, which represents a direct consequence of the historically changing and evolving objectively practical relationship between man and nature, a considerable portion of the creative activity of the cognizing subject is concentrated. This degree establishes the significance and orientation of the cognizing activity and simultaneously determine its actual limits and possibilities. The determination of the actual object of cognition is also variously mediated as a result of the contradictions given by the naturally historical character of cognizing activity. Mankind participates in the formation of the object of its cognizing activity with a certain degree of freedom and randomness which is, by and large, based on a certain amount of freedom in the approach to the realization of its own transforming practical activity in concrete historical conditions.

Scientific concepts contain elements of all procedures and proceedings with the help of which scientific cognition is carried out. From our point of view what is in this respect important is the role of what we call general scientific concepts. We have already mentioned that the epistemological value of a concept can be realized in many theoretical links. Scientific concepts which express a subject-object relationship in a very mediated manner may be activated by various theories both of a general and more specialized nature. This potential multidimensional connective nature enables the utilization of scientific concepts in the integrative process in science. In actual fact all scientific concepts possess this ability because they always constitute the unity of the general and the special. This ability, however, cannot be realized always and under any historically-cognitive circumstances for all standing scientific concepts but only for those concepts which comprehend the concrete historical requirements of the development of scientific cognition. And in a certain sense and in a certain concrete historical situation these concepts can be described as general scientific concepts, or better as concepts fulfilling the function of general scientific concepts. The integrative abilities of general scientific concepts ensues from the nature of the reflected objective reality, from the objective similarity of phenomena as a concrete manifestation of the material unity of the world, and from the subjective activity — through the grasping of these links — forms analogies in connexion with the historical possibilities and the requirements of the development of science.

From the above-mentioned preconditions then follows the basic possibility of utilizing the methods and concepts of any given scientific discipline as an instrument of cognition for another science. Such a possibility stems from the fact that sciences whose concepts assume the nature of general scientific concepts in the abstract level reflect relations which have not yet been revealed in their entirety in special sciences and with things being as they are, cannot be revealed either. General scientific concepts do not represent an immediate, direct reflection of objects studied by disciplines in which they operate in this general scientific role. But due to their nature, and if correctly mediated, they do become a means for mastering these phenomena. In their "general" aspects these concepts form a reflection of the objectively existing reality on such a level of abstraction and idealization which makes this mediation possible. The meaning of general scientific concepts for individual sciences is mediated. The rate and possibility of utilization of these concepts depend on the degree of development of the individual disciplines in that the formation of the general scientific framework makes the broader links and connexions of this discipline more precise.

In the above-mentioned sense general scientific concepts are "constituted" for example in connexion with the progressing mathematization of scientific disciplines, or generally speaking with formalization processes. The contributions made by mathematics, mathematical logic as well as cybernetics and other disciplines in this regard constitute a necessary factor for speading up the development of cognition in the various scientific branches in which a corresponding general theoretical basis has been laid down. In actual fact the mathematization of science is a very complicated process requiring an analysis of its own.

The general scientific and special scientific aspects of concepts are interconnected and this link is expressed within particular scientific disciplines too. The general scientific role of cognition can be significantly developed only in line with the development of special cognition which provides general scientific cognition with the necessary background without which a general scientific approach to reality cannot be attained.

The general scientific role of concepts in science is historically justified, as it expresses the qualitatively new nature of relations between individual spheres of knowledge, representing the implementation of the objective dialectical laws in the development of human cognition in the development of science. In addition it is a manifestation of the growing comprehensiveness of the entire scientific view together with deep penetration into the details of reality. At the same time, the fundamental determination of this possibility, i.e. the possibility of the growing comprehensiveness of cognition and knowledge, is established outside the epistemological sphere of cognition, being given by the growing complexity of the social transformation of reality — social practice, its current and perspective requirements. The integration in science as its inner process has a feedback to social practice.

The general scientific role of certain scientific concepts, as stated before, could originate and find their application in dependence on special scientific analyses of reality because only this prerequisite facilitated the revealing — with a corresponding precision — of the laws that in their abstract structure and in their general contents showed identical features captured by "general scientific" concepts. General scientific concepts combine general and special aspects of meaning; general scientific concepts on transformed into a concrete, special scientific background. General scientific concepts, however, are not immediately integrated in the system of concepts of a special science in which they play this role; but they are always used in a modified manner. This adaptation means that the given concept in its general scientific role is accepted in a corresponding sphere of abstraction and universality in which the particular scientific branch is based. At the same time, a network of links is created by means of which the general scientific concepts realize in this domain their theoretical and epistemological function. This process, however, posess certain dangers. First and foremost it is the danger of absolutization of a general scientific concept adapted on a certain level. This absolutization results in incorrect transformation of thus modified concepts immediately into other disciplines. Such utilization of a general scientific concept is both worthless and misleading from the epistemological viewpoint. At the same time, the process of creating and realizing the general scientific role of concepts is an inevitable methodological precondition for integration in science.

The integrative process forms an innermost process in the development of science, possessing its own objective laws and prerequisites. Not to respect them would amount to subjectivism, overestimation of one's own subjective efforts and interests and consequently to a methodological onesidedness. One of the manifestations of subjektivism is the incorrect introduction of concepts from a discipline which — due to its nature — can play a role in the integration in a given historical situation into other branches regardless of the possibilities existing in these disciplines as well as the established mediating and adaptating background. More attention has to be devoted in this respect to the application of general scientific concepts adapted in the sphere of natural science to social sciences. This process requires profound knowledge of the process of development of social sciences, the difference in their object of study and similarities of general laws common to natural and social reality, including their mutual links. On this basis, a methodology for the integration approach in social sciences and social, natural and eventually technical sciences can be elaborated.

If we proceed from the prerequisite that every scientific concept is part and parcel of the conceptual system of a particular scientific discipline, it follows that general scientific concepts, too, form parts of the conceptual structure of corresponding general scientific branches. The relationship between the systems of different general scientific disciplines is far from being simple. In every case it is impossible to transfer concepts from one system to another directly. The application of concepts of the general scientific system to a less general structure represents therefore a complicated process of "adaptation" and assimilation. Failure to understand this fact is often caused by a misunderstanding of the dialectical relation between the general and the special, by the confusion or identification of the general with the abstract, and the special — with the concrete in scientific cognition, which may lead, among other things, to a simplification or distortion of the cognizing value of general scientific concept. Moreover, and this is another very significant moment, the clasification of sciences into special and general scientific disciplines is relative in terms of content and history.

The general and the special always exist in a dialectical unity, like the abstract and the concrete, and their separation leads to onesidedness and distortions in scientific cognition too. The problem of the relationship between the general and the special on one hand and the abstract and the concrete on the other one, in connexion with the problem of the integrative role of scientific concepts, represents one of the most complicated issues.

This problem is closely connected with the problem of definition of general scientific concepts. Is it possible at all to present such a definition, for instance, of concepts such as "model", "system", "structure", "information" etc. which would be universally valid and applicable? The concept "model" can be found in modern science quite frequently, especially in links between various scientific branches. Its use is directly connected with disclosing similarities of phenomena and laws of different spheres of reality, both natural and social. The concept "model" has become one of the important means of expressing as well as utilizing cognized relations of similarity. Is it possible to give a general definition of the concept "model" to cover all these uses? In a formally abstract form in mathematics or mathematical logic and cybernetics there exists a definition of this concept. The concept "model", however, can also express such similarity relations which cannot be described in a formally abstract manner, or if it could, such a description would have no epistemological meaning. Relations of similarity contain not only a formal aspect but also a content aspect pertaining to corresponding properties, genetic and functional connexions, etc. In this sense "model", just as any other concept, can be defined from many aspects and in varying levels of abstraction and generalization. None of the definitions, however, has absolute validity. Nonetheless it is useful, and as far as methodology is concerned necessary, to seek and study interconnexions of the various uses of the concept "model" from the philosophical-epistemological point of view and according to this to analyze the epitemological-methodological meaning of the concept in question. In this sense it would be interesting for instance to study the use of the concept "model" in classical physics and mathematics. In mathematics the concept "model" occurs as a concept of this discipline itself, whereas in physics it appears rather as a concept having a methodological nature. At the same time, the mathematical concept "model" is a theoretical by abstract justification of the methodological function of model not only in physics but also in other special sciences, although within the limits of the possibilities offered by abstraction and generalization of mathematics.

As regards the integrative methodological function of general scientific concepts it is particularly important to analyse the link between general scientific concepts and concepts in philosophy. Scientific concepts, both in their special and general role, differ from philosophical concepts especially because of the differences in the object and function of these sciences and philosophy. Their relation is not merely a relationship between the general and the special, nor is it only characterized by a certain level of abstraction, even though these links, are inevitably present. It stems from the fact that the very relation between philosophy and special sciences is much more complicated and in many aspects more mediated. This mediation occurs not only on the level of cognition but by no means primarely. This fact is based on the objective reality and the practical requirements of society projected in various transformations into various links between special and general sciences and philosophy.

In their general scientific role scientific concepts are in a sense very close to philosophical concepts and categories. According to cases, sometimes philosophical and general scientific aspects of scientific concepts replace each other. This substitution is not — in our own view — permissible, since it violates or disregards the broader context of any particular concept. Such concepts like "model", "system", "structure", etc. are sometimes understood as philosophical concepts and are without any additional connexions used in philosophy. These concepts have, in fact, no immediate philosophical meaning. What then is the criterion for distinguishing philosophical concepts, especially philosophical categories, from scientific concepts in their general scientific role? Usually this criterion is sought in the degree of universality. We see the unreliability of this criterion in the fact that philosophical universality, i.e. the universality of philosophical concepts and categories, differs from the universality of other scientific concepts and categories, including their general scientific nature, not only quantitatively, but also qualitatively. In other words, philosophical universality is a something qualitatively different. Moreover, the relation between philosophical concepts and concepts of other sciences, including "general scientific" ones exists not only because of the relationship between the general and the special but also because of the relation between the special and the special. This stems particularly from the fact that the function of philosophy is not merely the generalization of the results of other sciences. The generalized experience of the special scientific general scientific investigations of reality represents a necessary prerequisite for solving philosophical questions. It forms part of the climate within which philosophical questions are raised and solved. It is one of the preconditions but certainly not a sufficient one; among the determining factors it is only secondary, behind prerequisites ensuing from the nature of the subject-matter and the social function of philosophy — i.e. to participate in the practical transformation of reality.

Concepts and categories of philosophy are results of generalizations of the entire cognizing and practical experience in the process of social development. The substantiality of individual philosophical concepts and categories stems from the specific character of philosophy as a science of the most general laws of the movement of nature, society and thinking. We subscribe to the view that the basic criterion of the differences in philosophical concepts and categories is their immediate link-up to the solution of the basic question of philosophy concerning the relationship between matter and consciousness, between the objective and the subjective.[3] In that sense no single philosophical concept, no single philosophical category can remain indifferent to its system as whole and thus to the fundamental ideological clash between materialism and idealism. This fact should be taken into consideration when tackling any seemingly special philosophical problems and when utilizing and interoducing new concepts into its system. It is understandable that new concepts appear and new categories are formed in the conceptual system of Marxist-Leninist philosophy due to the development and promotion of given epistemological and social preconditions.

Philosophical concepts and categories cannot therefore play the role of "indifferent" general scientific concepts. The integrative role of philosophy and, consequently, the integrative role of its concepts and categories, lies in its function as a world conception, including its general-methodological imports as well.

REFERENCES

1 V. I. Lenin, *Collected Works*, vol. 38, Moscow 1961, p. 213.
2 V. I. Lenin, l.c., p. 212.
3 V. L. Gott, A. D. Ursul, *Obshchenauchnye ponyatiya i ikh rol v poznanii* (General Scientific Concepts and their Role in Cognition), Moscow 1975, p. 24.

11. Contemporary System Concepts and the Integration of Scientific Knowledge

JANA PURŠOVÁ

11.1 INTRODUCTION

The explosion of new, significant theoretical facts in natural, technical and social sciences in the recent decades of our century enables us to speak of revolutionary changes in contemporary science. This situation necessarily calls for increased philosophical concentration on these highly topical questions. Enhanced interaction between philosophy and the other spheres of science may yield positive results.

One of the most characteristic features of theoretical knowledge has been, for some time, the inclination to a system interpretation of reality. These concepts react to the increasing differentiation which is taking place in all fields of science, influenced by the process of the scientific and technical revolution which, at the same time, makes it necessary to deliberate on the ways to integrate scientific knowledge. In this connection all scientific disciplines are clarifying their subject, methods, problems and targets.

In this respect, decisive importance is being ascribed to the term system[1] which reflects the general relationship of all phenomena and processes and is expected to facilitate the synthesis of scientific knowledge and the creation of a new image of the world. However, there are a number of unclarified points in this problem as a whole. In natural, technical and social sciences there exists a large number of various concepts of system, and even within the individual theoretical regions the terminology is still not uniform and the problem is open to discussion. In spite of all the differences, however, it is possible to distinguish some of the fundamental features which are common to the contemporary system trends in scientific knowledge. On the one hand, this involves a tendency towards a high degree of generalization and abstraction, on the other, a preference for methodological aspects with a frequent wide application of a number of formalized mathematical, logical and cybernetic methods.

The lack of understanding of the system explanation of nature and society, and the multiformity of these developing problems create difficulties in determining the rules for classifying this concept. Evidence of this situation is in the literature which frequently fails to distinguish between the individual trends and their theoretical and methodological orientation. We then refer to "system approaches", "system-structural approaches", "systems theories", "general theories of systems", "systems research", "system orientation", "systems philosophy", and a number of other attributes are associated with the term system. This terminological disunity then influences the interpretation of their content, they are frequently identified and a definition of their specificity cannot be reached.

The term system evidently has a different meaning, content and scope in the various fields and stages of scientific cognition. It is my opinion that the classification of systems concepts should be based on the varying degree of effort for generalization and abstraction in the individual sciences. From this point of view we can distinguish among several fundamental branches, featuring specific system characteristics of studying reality: intra-disciplinary, inter-disciplinary and supra-disciplinary systems concepts in natural, technical and social sciences, and trans-disciplinary system concepts in philosophy[2] which are then subject to further differentiation within the scope of these divisions[3].

11.2 INTRA-DISCIPLINARY SYSTEM CONCEPTS

Among these concepts belong the concepts of systems developed by the individual natural sciences such as theoretical astronomy, theoretical physics, theoretical chemistry, theoretical biology, etc., technology, as well as social science theories, e.g., sociological, economic, psychological, etc. Depending on their orientation, these sciences study actual inorganic, organic, technical and social systems and generalize the empirical facts thus acquired.

The purpose is to integrate the scientific facts of the individual disciplines of the same science without exceeding the scope of its own, either physical, chemical, biological, sociological, or other sphere; this is an integration which is stimulated by a profound internal differentiation of the individual scientific disciplines. I shall attempt to explain this problem on a few examples.

In theoretical *astronomy*, particularly in the field of cosmology[4], a number of remarkable opinions have appeared, related to the study of the Universe as a "system of all systems". Evolutionary cosmology emphasizes the hiearchism of material cosmic systems. It studies the multi-form motions of matter and seeks to determine the regularities of its evolution with the purpose of synthesizing the processes taking place in the mega-world, macro-world and micro-world. It is a very prospective science which, in co-operation with other sciences, may contribute significantly to our knowledge of the material unity of the world in its infinite diversity. The continued research in this field and its results are of considerable importance to world outlook.

In theoretical *physics* there is a number of important partial physical theories, united exclusively by the object of research, i.e. by the individual sphere of inorganic nature which exist or develop parallel to one another, but essentially independently of one another. They are achieving marked success and represent a substantial contribution to understanding the material unity of the world. However, they mostly deal with research into and description of the motion of micro-objects, which are understood as relatively stable systems, and their relation to the hierarchy of the other physical systems, to micro-systems, macro-systems and mega-systems, characterizing the objective reality as a whole, is not defined.

Contemporary theoretical physics thus deals with partial systems and processes of interaction as parts of a whole and reveals the partial regularities of these processes, however, it has not so far been able to understand the more general principles.[5] Hypothetically the world is considered to be a system, a whole, however, its character is not being sufficiently explained theoretically.

By understanding and unifying the multi-form regularities of the micro-world, macro-world and mega-world we shall apparently arrive at a uniform, internally consistent, general physical theory which will enable the inorganic world to be characterized as a whole. However, this theory should include not only the structural and functional, but also developmental aspects[6] in unity. We would then be able to begin explaining the relations between inorganic and organic nature as well as deliberate the consequences of the individual theories for social relations. The first steps towards integrating scientific knowledge have been made, however, comprehensive results will apparently only be achieved in the future. Theoretical physics are faced with the categoric imperative of a new synthesis.

The situation is similar in contemporary theoretical *chemistry* which studies the structural and functional aspects of processes taking place in micro-systems, various degrees of compounding and transmutation of atoms and various types of reactions. The purpose is clearly a higher synthesis of inorganic and organic factors, tending towards a uniform view of these processes, based on the summary of the latest knowledge in chemistry, biology and physics. Studying chemical systems capable of self-organization seems to be prospective and should lead to the establishing of a new conceptual apparatus of chemistry, to constituting evolutionary chemistry, establishing the theory of chemical evolution, which may also help in solving the problem of biogenesis.

In theoretical *biology* the role of evolutionary theories is now becoming more important[7]. Evolutionary biology deals with problems of the genesis and evolution of organisms in their interaction with the medium. It is characterized by a strong tendency towards integrating the facts known about organic nature and is attempting to create a new biological image of the world.

All biological objects are understood to be internally differentiated and, simultaneously, integrated systems in the hierarchy of natural systems. Concepts of macro-systems and mega-systems, exceeding the scope of the organisms and including the biosphere as a whole, are particularly important. In this way biology is reacting to the situation created at the beginning of our century, when the principles of studying macro-systems and macro-processes evoked by Darwin's evolutionary theory were discarded and orientation towards the study of processes within micro-systems began to prevail.

The methodological role of the evolutionary theory is emphasized. Considerable progress in this field of research can be observed in the dialectic concepts, particularly in Soviet science. This theory enables the knowledge of all partial disciplines to be combined and it seems that in this respect it is prospective. It is referred to as the synthetic theory of evolution which presents an idea of the structure, function and regularities of the historical evolution of the systems of organic nature.[8] The other biological theories only allow for the explanation of partial aspects of evolution.

A new item in the contemporary concepts of evolution is its approximation to antropological, sociological and ecological approaches, leading to attempts of a comprehensive view of the relation of man, nature, society and the environment.

The part played by theoretical biology in the process of scientific cognition is becoming more and more important. It is emphasized that this involves the process of dialectization of contemporary biology, expressed particularly by the enhancement of the integrational tendencies in the development of the biological sciences, by the increasing importance of the general branches of biology, strengthening of the connections among biology, physics and chemistry, the approximation of the structurally functional and evolutionary approaches, the weakening of the organismo-centric and elementaristic concepts, which, methodologically, leads to a change in the style of thinking. Mentions of the change in the style of thinking and of the contribution of biology to the dialectization and integration of scientific knowledge lead to hypotheses of the biological stage of the contemporary scientific and technological revolution, of the era of biology, which is stimulated by the revolutionary changes in this science.[9]

Technical theories, particularly in the field of systems engineering, assume a special position in science. They study technical systems, their structure, function, processes of transformation of substances, conversions of energy and the information processes which take place in them. These conclusions lead to creating new technical theories, on the basis of which other improved technical systems are then constructed. They are characterized by the interaction of natural, technical and social components and, from this point of view, they form the intermediary link between natural and social sciences.

In the *social sciences* systems concepts of social processes are now being widely developed. The Marxist science is developing the heritage of the classics of Marxism-Leninism in a creative manner and its findings are being integrated into a comprehensive system. In the bourgeois science the situation is more complicated due to the existence of a number of heterogeneous theoretical approches to the problems of the social system.[10] The tendency towards integrating socio-scientific knowledge is rare.

The Marxist-Leninist *sociological* theory, in its interpretation of society as a material system, is based on the fundamental concept of a socio-economic formation which combines the categorial apparatus as a whole and thus arrives at adequate theoretical conclusions. In the bourgeois sociological theory very general and abstract models of the social system are mostly formed.

In the *economic* theories the fundamental question in elaborating a systems concept is the approach to solving the problem of the theory of value, either objective or subjective, characteristic especially for various non-Marxist concepts. The adopted theory of value is then the element which holds the whole superstructure of the appropriate system economic theory together.

The brief mention of some of the problems of the individual sciences indicates that they will continue in improving the system characteristic of their subject. However, at the same time they will have to exceed the scopes of their own research and develop intensive inter-disciplinary co-operation, in order to arrive at an adequate system interpretation of reality and to apprehend the relations among the inorganic, organic and social spheres.

It will be necessary to integrate the results of scientific research of natural, technical and social sciences. This tendency is now being partly applied to the synthesis of scientific knowledge in some scientific disciplines, however, it is necessary to support it and develop it further.

11.3 INTER-DISCIPLINARY SYSTEM CONCEPTS

The philosophical problems created by enhanced interaction, mutual influence and the synthesis of the findings of the individual sciences have come to the fore quite distinctly in the recent decades. Many of these questions provide numerous stimuli for deliberation and discussion, because they represent a revolutionary factor in the evolution of scientific cognition.

This situation is also reflected in the terminology used. In the literature these disciplines are mostly characterized as "marginal", "boundary" or inter-disciplinary, sometimes also as "hybrid"[11] branches of science.

This terminology does not, however, reflect the essence of these disciplines in full. The individual disciplines are frequently co-ordinate mechanically with one another, insufficient account being taken of their subjects and contents. It is very difficult to adopt a more appropriate term, however, it would be more suitable to refer to them as *complex boundary disciplines,* in order to express better the aspect of interaction, integration, hierarchism and complex methods, used in studying the given region of phenomena with particular emphasis on the sphere which observes the higher forms of motion of matter.

The systems concepts, created within this framework in natural, technical and social sciences, deal with processes which take place in actual material inorganic, organic and social systems with emphasis on transitions and relations between the individual forms of the motion of matter, the purpose being to integrate the knowledge of the animate and inanimate nature, as well as of society, respecting the specificity of the individual systems.

The complex boundary disciplines substantiate the material unity of the organic, inorganic and social world. The sharp boundaries between the appropriate scientific disciplines, as well as between the individual forms of motion of matter are becoming less distinct. In accordance with this intensively developing trend in the evolution of scientific cognition, Engels' classification of the forms of motion of matter is being expanded.

The interaction and integration of scientific theories is taking place at several levels:

1. between the individual disciplines of the natural sciences (biophysics, biochemistry, physical chemistry,[12] nuclear chemistry, geobotany, geochemistry, geophysics, biomathematics, etc.);
between the individual disciplines of the technical sciences (e.g., power engineering);
between the individual disciplines of the social sciences (social psychology, economic history, etc.);

2. between natural and technical sciences (e.g., bionics, biocybernetics);
between natural and social sciences (historical geography, economic geography, social biology, etc.);
between technical and social sciences (social engineering, etc.);

3. between natural, technical and social sciences (medicine, ecology).

A great asset for the advancement of scientific knowledge is particularly the interaction of physics, chemistry and biology; we shall, therefore, attempt to present brief characteristics of biophysics, biochemistry, bionics and ecology.

Biophysics study the physical and physico-chemical processes which take place in organic systems, some of the partial physical manifestations of live matter, e.g., thermodynamic processes, energetic processes, acceptance and processing of information and other problems by means of biophysical methods.[13] According to this concept, the fundamental feature of the organism is constant exchange of substances, energy and information with the environment.

This theory, founded on the model of the biological system, is restricted, like all more recent theories, to the study and interpretation of the structural and functional aspects of concrete systems and does not consider the problems of evolution. Its subject is, in the first place, the capability of the organic system to accept and process information. It is exploited practically in solving technical problems, occurring particularly in the course of various manufacturing processes.

Theoretical concepts are also being developed intesively in *biochemistry*[14] which deals with the physiological processes of a chemical nature, taking place in live matter particularly from the point of view of the structural and functional aspects of metabolism. The name physiological chemistry would perhaps be more appropriate. It seems that the system nature of some biological phenomena could be characterized comprehensively by methods of molecular biology, which has arrived at the concept of biopolymeres, in co-operation with evolutionary biochemistry, in which the genetic method is beginning to gain ground.

One of the youngest disciplines, *bionics*, which combines some of the aspects of biology and technology, studies the structure and function of the biological system as a technical object by means of a physical apparatus. It concentrates, in particular, on problems of self-regulation, as well as on the communication and processing of information in organic systems. The created models are very useful in technology and medicine. From this point of view bionics come close to cybernetic models and theories.[15] Bionics are considered to be a technical science which studies the structural and functional analogies in technical and organic systems.

Recently, a trend has begun to appear for the combining of biological and sociological theories[16] on a new theoretical level, directed at achieving a complex, system understanding of man and at a more profound explanation of the dialectics of the relation between the biological and social aspect which does not, however, proceed in the foot-steps of the out-dated social Darwinism.

Ecology now has a special status as an independent science, which combines theoretical findings about eco-systems of practically all natural, technical and social sciences and which uses their specific research methods. It is founded on the co-operation among the spheres of geology, geography, physics, chemistry, biology, medicine, technology, sociology, psychology, economics and of a number of others. It deals especially with global problems of the dialectic relation of man and other organisms to the natural enviroment, with the problems of the relations among organisms, and problems of interaction of nature and society. Ecology is a prospective science, because it is capable of influencing rationally the fate of our planet and the future of mankind.

The role played by *historical ecology*, which will deal with the historical evolution of the enviroment of man and of human society, special and general ecology, will apparently also become more important.

The question of the interconnection of the sciences of organic and inorganic nature and the society, particularly of the relation between biological, chemical and physical systems and processes, still remains open. There appear tendencies towards unilateral mathematization, physicalization, cybernetization or biologization of all scientific knowledge, which will clearly not provide an adequate way to integration. The necessity of a complex synthesis of not only natural-science and technical, but also social knowledge, which must be combined, is being forgotten.

In this respect, the problem most frequently discussed is the application of physical theories and models to chemistry and biology. The opinion prevails that physical theories are capable of significant contribution to explaining the individual relations among these systems and processes, provided they avoid reducing the higher forms of motion of matter to lower. This approach may yield a number of positive results and determine some of the relations between the physical, chemical and biological spheres,[17] and their exploitation in the field of technology and in social practice. However, in some non-Marxist concepts this complicated problem is being mechanically simplified.

The role played by natural-science theories in the proposed integration of scientific knowledge is being discussed frequently. Individual authors mostly prefer their own idsciplines. Decisive importance, in this connection, is erroneously being ascribed exclusively to physical theories or biology. At other times cybernetics are considered to be the integrative science. Important results are also being expected from the unification of the principles of cybernetics, the systems theory, the theory of information and mathematics[18].

The conclusion is also being reached that the contemporary scientific and technological revolution and, in particular, automation has enabled the transformation of the physico-chemical science of the 1st half of the 20th century into the cybernetico-biological science of the second half of this century.[19] The requirement for dialectic synthesis of physics and biology has also appeared, the combining of the conceptual bases of both fields being an important integration factor of scientific knowledge and a condition for creating a new scientific image of the world.[20] The synthesis of various inorganic, organic and social principles is also being considered. For example, the synthesizing nature of ecology is being emphasized. The concept of a socio-bio-geo-system, which may qualify for a means of integrating the knowledge of social and natural-sciences, has also appeared.[21]

The principle of the material unity of the world, substantiated by the transitions between the individual forms of motion of matter, implies that the systems interpretation of objective reality requires extensive and intensive co-operation among the individual fields of science, which will also contribute to the concretization of the conclusions of the individual disciplines. The interdisciplinary approaches to these problems are very fruitful and prospective.

In defining the status and role of inter-disciplinary systems concepts it is necessary to emphasize that they are a concrete, special and very effective form of integrating scientific knowledge.

It is my opinion that achieving a complex characteristic of objective reality requires a harmonization of intra-disciplinary research, which represents the first stage in system cognition and solves the theoretical and methodological problems of one discipline with an inter-disciplinary approach. Natural-science theories are capable of providing an adequate system explanation of reality, because they are able to verify each achieved degree of knowledge experimentally, as well as by practical application of its conclusion to actual organic and inorganic systems.

Both the system trends mentioned, characterized as intra-disciplinary and inter-disciplinary, have a number of common features. They formulate the system principles of establishing their own theory, in which they sometimes also include the theoretical findings of related disciplines. Their important specific feature is that abstraction and generality are not the initial factor, but the result of studying all concrete physical, chemical, biological, technical and social systems, existing in organic and inorganic nature and in society. The endeavours to integrate the knowledge of natural, technical and social sciences will clearly be intensified in the future.

11.4 SUPRA-DISCIPLINARY SYSTEM CONCEPTS

In other trends of contemporary scientific thinking, in general, supra-disciplinary systems concepts, a different approach is being adopted. The endeavour to build up a concept of systems, founded on general principles which are to be valid for all sphere of reality, predominates.

These concepts avoid the qualitative differentation of concrete systems and inorganic, organic and social processes, and they seek to determine the most general relations valid for the various forms of motion of matter. From generalization and abstraction, however, they proceed to the application of the system principles to actual objects with the purpose of verifying and improving the accuracy of their original premisses. In most cases methodological aspects are greatly preferred.

The supra-disciplinary systems concepts now form a relatively independent trend of thinking, concentrated on the problems of complex dynamic systems. In this respect one even refers to system philosophy, system orientation, system era, system image of the world, system opinion of the world, etc. The fact that contemporary system theories contribute to the dialectization and mathematization of scientific knowledge, is frequently emphasized.

We are able to distinguish between three of the strongest trends of general system thinking: the general systems theory, cybernetically oriented theories of systems and system approaches. The initiator of the general system theory was L. von Bertalanffy, whose concept has been developed extensively in contemporary American science by K. E. Boulding, R. W. Gerard, A. Rapoport and a number of others. Systems approaches have been applied in contemporary Soviet science in the publications of I. V. Blauberg, E. G. Yudin, V. N. Sadovsky and also of other authors. The general systems theories also include cybernetic systems concepts, inspired by N. Wiener, which are now being advocated by a number of Marxist and non-Marxist scientists. Recently, the synthesis of the theoretical principles of the general theory of systems and of the cybernetic theories of systems is being considered.

The *general system theory* developed by L. von Bertalanffy since the thirties of this century, was originally based on biology, however, it gradually crystallized into an increasingly general form of synthesis. In the first stage of his work Bertalanffy drew on the existence of actual biological systems and in this stage he referred to his theoretical base as "organicism". His theory acquired increasingly abstract features and tended towards creating a universal theory of sciences. Later, a formally logic and mathematical apparatus began to be formed for this field (Rapoport, Mesarović).

Bertalanffy frequently emphasized that his original contribution was in his elaborating his own version of the general theory of systems even before cybernetics and systems technology existed. The most important component of this concept, according to him, is the general theory of the hierarchic order, encompassing the structurally functional point of view, the principles of which can be formulated verbally. A number of terms, according to Bertalanffy, can widely be applied to material, psychological and socio-cultural systems.[22] He also pointed out the meta-scientific and philosophical aspects of this theory.

In one of his last works he emphasized the necessity to achieve the synthesis of scientific knowledge and recapitaluated his concept of the world as an organization which is to evoke a new integration and conceptual unification of contemporary science.[23] Bertalanffy gave three sources of modern systems research, i.e. his own concept, cybernetics and the requirements of technology with regard to controlling complicated manufacturing processes. He distinguished between two trends in the contemporary evolution of sciences: mechanistic, represented by cybernetics, the theory of automatons, computers, etc., and organicistic, represented by the general system theory. The new organicistic world opinion, according to Bertalanffy, is simultaneously "prospectivistic", because it is aware of its limitations, determined by the limits of the nature and culture of man.[24]

The other representative of the general systems theory, A. Rapoport, also emphasized that the task of this theory was to unify the individual branches of science, the regularities observed within limited regions, being understood as partial cases of more general regularities.

He based his method on biology, technology and mathematics. The contribution of biology, according to his opinion, is in that the term organism is applied to a wide range of phenomena, including social organism, and, consequently, it is reasonable to look for a generalization of biological laws which would lead to a general system of the organic type. The contribution of technology is in applying the term machine to complicated systems, capable of self-regulation, which simulate live organisms.[25] Inspired by cybernetics, Rapoport is an advocate of unifying the theoretical principles of biology and technical disciplines.

For example, E. Laszlo, who follows in the footsteps of Bertalanffy, also professes the general systems theory; he justifies the necessity to develop the "systems philosophy",[26] which is to reflect the ontological essence of all existing systems, natural and social alike, with the aid of the method of creative deduction.

Versions of the general systems theory, modified in various ways, attempt to solve the demanding philosophical problems to which, however, they are incapable of providing an adequate solution due to their unilateral approach. These "meta-scientific" abstract deliberations do not reflect the essence of the actual inorganic, organic and social processes, and it is problematic whether they are capable of creating a "new universal theory of science".

Since their birth in the middle of the 20th century, *cybernetics* have been attempting to synthesize the general principles and regularities of the functioning of the individual spheres of reality. Like the general systems theory it establishes a certain abstraction of the system and only them goes on to apply these principles to actual natural and social systems. It studies the structurally functional regularities of motion of complicated self-regulating systems, the regularities of communication with emphasis on obtaining, transfer and storage of information, as well as questions of management and control, and other functional aspects.

The basis is the analogy among inorganic, organic and social systems, for which a functional, highly organized, regulated exchange of substances or energies and information, preserving the stability of the structure within the limits of certain parameters and decelerating the growth of the entropy of systems, is characteristic. An important part is played by cybernetic models which, with the aid of electronic and other equipment, characterized by the capability of imitating some functions of organic systems, enable us to discover some of the principles of complicated processes and to determine the differences between the motion of natural, organic and man-made, technical systems.

One of the Marxists authors defined cybernetics as the theory of relations of dynamic self-regulating systems to their partial systems. According to him, cybernetics are dynamic and dialectic. They are founded on the principle of mathematically expressing the behaviour of systems in biology, technology, physics, the social sphere as well as in other regions.[27]

Cybernetics work with the general concept of special types of systems which organize, control and regulate themselves. However, there is the danger that the principal qualitative differences among the individual concrete types of systems will be underestimated.[28] Some non-Marxist theories, which are founded on erroneous premisses and identify the regularities of organic and technical systems, are based on the incorrect interpretation of the principles of cybernetics.[29]

Certain structural and functional analogies between organisms, machines and society indubitably exist. However, the essence of various types of these systems rests in the specific regularities of the inorganic sphere, of the higher organic sphere and of the most complicated social sphere. This peculiarity is frequently underestimated in some interpretations of cybernetics and the problem is simplified by a reductionistic approach.

Partial associations cannot be upgraded to a universal principle, the key to interpreting reality as a whole, as is frequently the case in some theories. It is, therefore, impossible to determine the regularities valid for all systems only on the basis of cybernetics; only partial regularities for partial sphere can thus be explained, the peculiarities of the individual forms of the motion of matter being respected.

Cybernetics are very important in studying all systems. The cybernetic principles, which were stimulated by the increasing importance of automatic processes, are an asset not only to the development of science, but also play a significant role in practical life, in technology, economy, in management, etc. Cybernetics have achieved very substantial successes, but one must realize at the same time what its place and role in scientific cognition is, because the validity of cybernetic principles is limited. In relation to philosophy[30], cybernetics are rather an auxiliary science, a means to understanding the important partial aspects of reality, capable of contributing to the formation of the scientific world outlook, but unable to replace it.

In this connection we should draw attention to the special status and integration properties of *mathematics* and *logics*. These sciences deal with general structural relations, valid for ideal systems and their sets, abstracting from the actual properties of systems, like cybernetics. The mathematical models of these objects are methodologically characterized by a high degree of formalization.These scientific fields are an important means of understanding objective reality. Mathematization and formalization are an important support for other scientific disciplines. Their conclusions contribute considerably to philosophy and to the dialectic generalization and forming of the scientific world outlook. However, once again we must avoid making mathematical and logical principles absolute.

A special chapter in the tendencies to the general systems characteristic of objective relity is formed by *system approaches* and the *general system theory* which have recently begun to be developed in Soviet science. Although they are a relatively young branch of the contemporary scientific knowledge, the literature concerning it is quite abundant.[31]

The mentioned *general system concepts of a supra-disciplinary nature* are intended to integrate scientific knowledge by way of generalization and abstraction, which enable the general principles which are transferred to the individual types of inorganic, organic and social systems, to be determined. The basis for understanding the actual systems is, therefore, the creation of an abstraction of the system, which cannot reflect reality as a whole. Most conclusions are only valid for partial spheres in the fields of theory and practice alike.

In studying objective reality as a system we find it necessary to identify the gnoseological and methodological aspect with the world-outlook aspect. It is also necessary to take into account the mutual percolation of the methods, because the theory acts on the adequate selection of methods and the development of methods conversely on the region of theory, etc. The methodological aspect is frequently overestimated particularly in cybernetics and in systems approaches. The general theory of systems again gives preference to the ontological aspect in detriment of the methodological.

A very controversial and frequently discussed problem is the definition of the status and task of the general systems theories in scientific cognition. The American version of the general systems theory claims world-opinion validity of its conclusions. The "systems" methodology is frequently even raised to the rank of a philosophical methodology, but at other times, on the contrary, the theoretical conclusions lack an adequate methodological foundation. In some cases no difference is made between Marxist and non-Marxist concepts of the system and this leads to the convergence of the dialectic and metaphysical approach to the problem being investigated.

We frequently hear of a system structural approach or method. One has to consider these terms which are closely associated with conceptual approaches. If we speak of a system, we should be referring to the unity of structural, functional and evolutionary aspects, the synthesis of which characterizes the system as a whole. For a more detailed analysis this whole can be resolved into individual components, however, the study must always reach a certain higher form of synthesis and must not stop at the analytical stage.

Only by using a series of classical methods of materialistic dialectics is one able to arrive at a synthetic view of natural and social processes. From this point of view we must then necessarily consider the system-structural approach to be one of the analytical aspects of the overall gnoseological concept of the system, concentrating only on the study of the structural, or structural and functional aspects of given systems. However, provided the individual concepts include the structural, functional and developmenta aspects, it would be more suitable to refer to them as "systems theories", which is indicative of the nature of the basic origins.

The general systems theory of a supra-disciplinary character, however, mostly does not take the genetic and developmental aspects and its qualitative characteristics into account. Although they frequently advocate the existence of developmental relations, they do not analyse them and the structurally functional determinants and quantitative changes within the systems being studied remain in the limelight of attention.

11.5 EVOLUTIONARY TENDENCIES OF THE INTEGRATION OF SCIENTIFIC THEORIES

It is evident that efforts are being made in contemporary science to produce a "systems" image of the world and that there are different ways of arriving at this synthesis. The problems involved are very complicated, because they are affected considerably by world-outlook aspects and they must, therefore, be treated in relation to materialistic dialectics. Philosophy generalizes the results of concrete sciences, on the one hand, and forms the methodological base for the further development of scientific knowledge, on the other. It has a trans-disciplinary character. The condition for treating this problem successfully is close co-operation between natural scientists and philosophers.

If we deliberate the evolutionary tendencies of systems concepts, it would seem that the assets would primarily be the theoretical results of the actual natural, technical and social sciences and their interdisciplinary co-operation in the further evolution of science. The systems principles follow from the content, from the essence of the individual theoretical disciplines, which must clarify and verify their bases by specific means. A number of scientific disciplines has now reached a high standard; a large amount of new facts have been accumulated, however, synthesis, which requires time, is only being expected in a number of fields.

In this respect, Engels' opinion is very interesting, i.e. that the tremendous amount of facts accumulated by natural science requires them to be classified in each individual field systematically and according to their internal relations. He is also of the opinion that the individual fields of cognition must be interrelated.[32] This conclusion has a more general validity and it implies consequences not only for the organization of scientific research, but also for social practice.

It is clear that a reclassification will have to be made in the whole system of sciences, which will be in accordance with the corresponding approach to the scientific research into the individual forms of motion of matter, as well as to the transitional, complex boundary disciplines from the point of view of dialectical and historical materialism, and which will bring order to the divison of sciences, created by spontaneous evolution.

Data for further philosophical treatment of the results of the individual fields of science are gradually being accumulated. No intermediary links in the form of a theory and methodology of "intermediate range", located between the actual sciences and dialectical materialism which is capable of generalizing and synthesizing the facts about partial systems, of determining the general systems regularities and of expressing their general essence, will be necessary. This process will be accelerated by further penetration of dialectical principles into the individual spheres of science.

As regards the more general, supra-disciplinary systems concepts, particularly the results of cybernetics and mathematics will be important for the future of this process, provided they do not claim world-outlook validity for their fundamental principles. The cybernetic systems theories will be very useful in application to the various spheres of technology and economics, in studying a whole series of partial aspects of inorganic, organic and social systems. They are an important auxiliary means of dialectical analysis of reality. However, one must realize that their degree of generality is lower than that of philosophy, which provides for their adequate explanation and application.

Contemporary science and its individual regions are widely differentiated and specialized, which supports the efforts to create various supra-disciplinary theories which are intended to reflect the unity of the world, the principles and results of materialistic dialectics and its importance for integrating theoretical knowledge being frequently neglected.

Materialistic dialectics as the science of the most general regularities of the evolution of nature, society and thinking, which develops the heritage of the classics of Marxism-Leninism, plays a decisive part in forming and moulding the world opinion; this enables the scientific images of the world, produced by the individual sciences, to be combined. The principle of unity of world-outlook, epistemological, methodological and practical aspects is the condition for the adequate interpretation of the natural and social reality and the base for studying all systems, what follows from the trans-disciplinary character of philosophy.

The application of the method of proceeding from the abstract to the concrete and vice versa, of qualitative analysis as a way to synthesis and a number of other methods of dialetical materialism, together with the use of modern mathematical and cybernetic methods and their correct evaluation, leads to an adequate understanding of the essence of natural and social processes. It enables us to determine the regularities between the individual processes, the transitions of the individual forms of motion of matter and the regularities of this motion.

The problems, with which all the systems theories mentioned above deal, will gradually become more lucid as we proceed from the description of terms, which now mostly predominates, to the analysis of their content. The most topical task is clearly to explain the relation between the content of the categories, used in the concrete natural, technical and social sciences, and the content of the general categories of philosophy. This orientation will also permit a closer association of the logical-methodological and theoretical aspects. Unity, conceived in this manner, is the condition for understanding the systems problems in all sciences and will help overcome many terminological and conceptual difficulties as well.

The smoothest way to integrating scientific knowledge, to overcoming the present diversity of systems characteristics of objective reality and to moulding the scientific world-outlook is the profound and consistent deliberation and elaboration of the conclusions of dialectical and historical materialism. In this respect the following thesis of Engels is very topical: "... dialectics are indeed the most important form of thinking for present natural science, because only they provide an analogue and thus also a method of explaining the evolutionary processes, taking place in nature, the overall relations, the transitions from one field of study to another."[33]

This procedure may serve as a guide to a creative elaboration of the systems characteristic of reality, on the one hand, and a basis for evaluating contemporary systems concepts and approaches, and for assessing their contribution not only to the development of scientific knowledge, but also for social practice.

REFERENCES

[1] The term system played an important part in the history of modern science since the 16th century. However, the various versions of older concepts mostly only reflected a very limited region of phenomena and emphasis on the ontological aspect of the studied reality predominated in them. In connection with the change of the theoretical orientation of science in the 20 th century, the term system acquired a qualitatively new content. Its methodological importance began to be emphasized more. Moreover, it is evident that an adequate explanation and application of the term system can only be achieved by employing world-outlook, epistemological and methodological principles together.

[2] The terms intra-disciplinary, supra-disciplinary and trans-disciplinary are not used currently, however it is my opinion that in this brief review of the principal trends of contemporary system thinking they are expressive.

[3] I have already outlined some of the aspects of this concept earlier. Compare, e.g., with J. Puršová, ,,Pojetí systému v soudobé vědě" (The System Concept in Contemporary Science), in: international transactions *Dialektika a systémový přístup* (Dialectics and the System Approach), Praha 1979, pp. 190–202; J. Puršová, ,,Koncepce

systému a problém integrace přírodovědeckých teorií" (The System Concept and the Problem of Integration of Natural Sciences Theories), *Dějiny věd a techniky*, vol. 12, 1979, No 2, pp. 94—109.

4 A contribution not only in the field of cosmology, but also cosmogony and astrophysics are the investigations of V. A. Ambartsumyan, contained in several monographs.

5 This fact was pointed out, e.g., by H. Hörz, *Fyzika a světový názor* (Physics and World Outlook), Praha 1973, p. 88.

6 The question of evolution in inanimate nature is being discussed extensively in the literature, however, the problem has not been solved.

7 We distinguish between three fundamental types of evolutionary concepts in the 20 th century, i.e. the dialectical-materialist, the mechanistic and idealistic. Refer to V. M. Mednikov, *Darwinismus ve 20. století* (Darwinism in the 20th century), Praha 1980 p. 13.

8 The dialectical explanation of these aspects is the purpose of A. S. Mamzin's work *Ocherki metodologii evolyutsionnoy teorii* (Outlines of the Methodology of the Evolutionary Theory), Leningrad 1974, compare, e.g., p. 10.

9 Compare, e.g., I. T. Frolov, "O dialektike biologicheskogo poznaniya" (On the Dialectics of Biological Knowledge), in: *Filosofiya i estestvoznanie*, Moscow 1974, pp. 206—207; L. F. Ilyichev, *Filosofiya i nauchniy progress* (Philosophy and Scientific Progress), Moscow 1977, p. 139; R. Löther, *Biologie a světový názor* (Biology and World Outlook), Praha 1973, pp. 11—12.

10 I have dealt with these problems in my book *Vývoj systémových teorií společnosti*, Příspěvek ke kritice buržoazní filozofie a sociologie (Development of Systems Theories of the Society; A contribution to the criticism of bourgeois philosophy and sociology), Praha 1980, 490 pp., in press, Academia, Publishing House of the Czechoslovak Academy of Sciences.

11 See M. C. Chepikov, *The Integration of Science*, Moscow 1978, p. 27.

12 The literature on the problems of physical chemistry was enriched by the publication of the American textbook: W. J. Moore, *Physical Chemistry*, Engl. Cliffs, 1972; the Czech translation Fyzikální chemie, Praha 1981, 974 pp

13 This formulation specifies the opinion maintained, e.g., by W. Beier, *Biofyzika* (Biophysics), Praha 1974, p. 12, that biophysics studies live systems using general methods of physics and that it views the organism as a material system in space and time, which differs from the phenomena of inanimate nature in its higher degree of complexity.

14 One of the authors is of the opinion that "biochemistry is part of the science of life. Its task is to study the manifestations of life by means of chemical methods".— P. Karlson, *Základy biochemie* (Foundations of Biochemistry), Praha 1981, p. 15. Recently an extensive textbook of biochemistry was published. See D. E. Metzler, *Biochemistry, The Chemical Reactions of Living Cells*, New York, San Francisco, London 1977; Russian translation D. Metzler, *Biokhimiya, Khimicheskie reaktsii v zhivoy kletke*. Tom 1, 407 pp., Tom 2, 606 pp., Tom 3, 488 pp., Moscow 1980.

15 Identical features and differences between bionics and cybernetics are analysed by H. Heynert, *Úvod do obecné bioniky* (An Introduction to General Bionics), Praha 1977, pp. 32—33. This author does not consider bionics to be a boundary discipline, but an independent discipline of applied biology. Compare p. 36.

16 This orientation has recently been appearing in Soviet literature. In non-Marxist science this problem in another form occurs under the name biosociology, mostly in connections with the earlier biosocial trends in sociology, Compare, e.g., with

W. W. Spradlin, P. B. Porterfield, *Human Biosociology, From Cell to Culture,* New York 1979, 227 pp. There is the danger that in some versions the concept of man may become unilaterally biologized.

[17] The knowledge of all three scientific regions are combined in biophysical chemistry. Two important books have recently been published. Viz V. Kalous, Z. Pavlíček, *Biofyzikální chemie* (Biophysical Chemistry), Praha 1980, 350 pp.; and the book of the American author A. G. Marshall, *Biophysical Chemistry,* New York 1978, translated into Russian, A. Marshall, *Biofizicheskaya khimiya,* Printsipy, tekhnika i prilozheniya v 2-kh tomakh, Tom I 358 pp., Tom 2 820 pp., Moscow 1981.

[18] G. Klaus, *Rationalität—Integration—Information,* Berlin 1974, p. 12, 82—83, 97.

[19] S. N. Smirnov, *Dialektika otrazheniya i vzaimodeistviya v evolutsii materii* (Dialectics of Reflection and Interaction in the Evolution of Matter), Moscow 1974, p. 368.

[20] Compare with I. A. Akchurin, *Edinstvo estestvonauchnogo znaniya* (Unity of Natural-Scientific Knowledge), Moscow 1974, p. 7.

[21] I. B. Novik, "Budushcheye nauki i ekologicheskaya problema" (The Future of Science and the Ecological Problem), *Voprosi filosofiii,* 1975, No 1, pp. 114—115.

[22] L. von Bertalanffy, "General System Theory: Problems and Results" (In Russ), in: *Sistemnye issledovaniya,* Ezhegodnik 1969, Moscow 1969, pp. 49—50.

[23] L. von Bertalanffy, *Člověk — robot — myšlení* (Man—Robot—Thinking), Praha 1972, p. 90.

[24] Ibid., p. 139—140.

[25] A. Rapoport, "Razlichnye podkhody k obshchei teorii sistem" (Various Approaches to the General Systems Theory), *Sistemnye issledovaniya,* Ezhegodnik 1969, Moscow 1969, pp. 78—79.

[26] Compare with E. Laszlo, *Introduction to Systems Philosophy,* New York 1972, pp. 3—14.

[27] G. Klaus, *Kybernetik in philosophischer Sicht,* Berlin 1961, pp. 191—192.

[28] Compare with M. V. Mostepanenko, *Filosofiya i metody nauchnogo poznaniya,* Leningrad 1972, pp. 195 (Philosophy and Methods of Scientific Cognition).

[29] Compare, e.g., with W. Wieser, *Organismen, Strukturen, Maschinen,* Frankfurt a. M., p. 13; K. Steinbuch, *Automat und Mensch,* Berlin 1961, p. 10.

[30] The relation between cybernetics nad philosophy is analysed in detail by H. Hörz, *Marxistische Philosophie und Naturwissenschaften,* Berlin 1974, in particularly compare with pp. 153, 188.

[31] I shall not discuss in detail the theoretical and methodological bases of the representatives of this type, because they are contained in these collected works. The explicit expression of the problem and purpose of systems approaches and of the general theory of systems is contained in several volumes of *Sistemnye issledovaniya,* which have been published since 1969 in Moscow. Several more extensive works can also be mentioned: I. V. Blauberg, E. G. Yudin, *Stanovlenie i sushchnost sistemnogo podkhoda* (Determination and Essence of the Systems Approach), Moscow 1973, 270 pp.; V. N. Sadovsky, *Osnovaniya obshchei teorii sistem* (Foundations of the General Systems Theory), Moscow 1974, 279 pp.; I. V. Blauberg, V. N. Sadovsky, E. G. Yudin, *Systems Theory, Philosophical and Methodological Problems,* Moscow 1977, 318 pp.

[32] K. Marx, B. Engels, *Spisy* (Works), vol. 20, Praha 1966, p. 344.

[33] B. Engels, "Dialektika přírody" (Dialectics of Nature), in: K. Marx, B. Engels, *Spisy* (Works), vol. 20, Praha 1966, pp. 344—345.

12. Functions of Abstract Structures in the Integration and Mathematization of Scientific Knowledge

G. I. RUZAVIN

The concept of "abstract structure", first lucidly formulated in mathematics in the 1930s, plays an important role both in the foundations of mathematics itself and in the theoretical analysis of many essential processes in the development of scientific knowledge today.

The paper deals with problems in ascertaining the role and functions of abstract structures when mathematical methods are applied to concrete sciences, i.e. when mathematization of scientific knowledge takes place. On this basis it becomes possible to show also the integrative role of abstract structures in the development of both mathematics and, indirectly, other sciences in which mathematical methods are used.

12.1 THE CONCEPT OF "ABSTRACT STRUCTURE": HISTORICAL BACKGROUND

Before considering the present-day definition of the concept "abstract mathematical structure", it is necessary to outline how it emerged and developed.

It is commonly accepted that the concept in the way it is used in mathematics today appeared as a result of an interplay of ideas associated with the axiomatic method, on the one hand, and Cantor's theory of sets, on the other hand. Furthermore, the teachings and theories themselves were developing, as we know, under the influence of studies carried out in various branches of mathematics. Indeed, a great impetus to the development of teaching about the axiomatic method was given by the discovery of non-Euclidean geometry and the theory of sets appeared in response to the development and substantiation of mathematical analysis.

The inception and development of the concept "mathematical structure" show how an entirely new approach to mathematics and the essence of mathematical knowledge itself comes into being through the integration of individual mathematical theories, their generalization and development on the basis of an axiomatic method.

This approach traces its origin from the discovery of a new, non-Euclidean geometry by Lobachevsky and Bolyai. The theoretical-cognitive and general scientific importance of this discovery can hardly be overestimated. It not only undermined the thousand-year long belief in the existence of only one Euclidean geometry, but also changed drastically previous ideas of geometry and then mathematics as a whole. First, this discovery showed that different geometrical systems could be used in describing space properties and, hence, the systems themselves are abstract structures rather than empirical and inductive generalizations of facts. Secondly, it is only experience and the astronomical and geodetic measurements that N. I. Lobachevsky and K. F. Gauss used to verify that the new geometry can prove which of the abstract systems of geometry better suits the description of reality. Thirdly, by interpreting abstract geometrical systems it became possible to clearly establish differences between purely mathematical structures and their applications.

Consequently, the axiomatic method itself underwent significant changes. Formerly axioms were considered certain obvious truths about space properties; later, when E. Beltrami offered a feasible interpretation of non-Euclidean geometries with the help of objects of Euclidean geometry, B. Riemann and then other scientists began to consider hypotheses whose applicability to the study of special properties and relations needs to be proved following the appropriate interpretation of basic geometrical concepts. The concepts themselves, even though they continued to be called point, straight line, plane, etc., proved to be as abstract as algebraic formulas themselves within a purely mathematical study.

This view of geometry was developing gradually and took its relatively completed form by the end of the 19th century in the book of D. Hilbert, the famous German mathematician.

He wrote: "We conceive three different systems of things, things of the first system we call 'points' and mark them $A, B, C, \ldots$, things of the second system we call 'straight lines' and mark them $a, b, c, \ldots$, things of the third system we call 'planes' and mark them $\alpha, \beta, \gamma \ldots$ We conceive points, stright lines and planes in certain relationships among themselves and denote these relationships using such words as 'lie', 'betwen', 'congruent', 'parallel', 'continuous'. The exact and mathematically comprehensive description of these relationships can be achieved using geometrical axioms."[1]

The important new step towards the development of the axiomatic method, which is sometimes called a revolution in axiomatics, is, therefore, associated with the transition from the concrete or material axiomatics, of which Euclidean geometry is an example, to the abstract axiomatics of the Hilbert type. Later, to analyze problems in establishing the basis of mathematics, scientists began to use semi-formalized and formalized axiomatic systems employing symbols instead of terms and formulas instead of propositions.

In completely formalized axiomatic systems even propositions or inferences came to be regarded as transformations of one formulae into others using rigid transformation or deduction rules. The results of such reasoning appear as *calculus*.

The axiomatic method makes it possible to formulate common ideas and concepts of mathematics, provided they are revealed by a creative generalization of the most important ideas and theories. Such new ideas, therefore, cannot help affecting existing mathematical knowledge. Time and again a need is felt in mathematics to reconsider existing knowledge from a common viewpoint. Such a viewpoint materializes itself in a fundamental theory which is regarded as the basis of mathematics for a certain period. Beginning with the 1870s, the abstract theory of sets proposed by G. Cantor moved in to the foreground.

The general and abstract character of the set theory already becomes evident from the fact that it treats such objects of all mathematical theories as numbers, functions, matrices, vectors and operators, irrespective of their meanings, i.e. morely as elements of the infinite sets than can be constructed using certain rules. An attractive trait of the set theory is that it actualized the concept "infinity" as a result of which the infinite appears to a degree as the finite. This facilitates considerably the application of elementary laws of classical logic to reasoning about infinity. Since mathematics deals, above all, with infinite sets of objects, it is only natural that this approach soon merited recognition among scientists and the set theory itself was treated as the unshakeable foundation of mathematics up to the time when several paradoxes were found in it.

The concept of abstract structure existed in implicit form as far back as the early 20th century, but explicitly it was formulated as late as the 1930s by a gifted group of young French mathematicians writing under the pseudonym N. Bourbaki.

In their programme paper entitled "Architecture of Mathematics", abstract structure was defined as follows: "To determine the structure it is necessary to specify one or more relations involving its elements... then to postulate that a given relation or relations meet certain pre-set conditions that are *axioms* of the structure in question. To build up an axiomatic theory of a given structure means to deduce a logical inference from the structure's axioms, relinguishing any other suppositions concerning such elements, specifically relinguishing any hypotheses about their origin."[2] It is apparent from the definition that all mathematical objects in such reasoning can be treated as elements of abstract structures, all the necessary properties of which being specified by axioms. Depending on the nature of these properties, Bourbaki singles out three basic or generating *types* of structures: algebraic, topological, and ordinal.

The concept of abstract structures makes it possible both to understand and actually to build up mathematics or at least most of its parts by integrating or unifying generating structures. The complex structures of this kind, as N. Bourbaki stresses, are not only the structures "superposed upon one another", but "intrinsically combined with one or more axioms connecting the structures".[3]

The integrating role of the theory of abstract structures manifests itself, firstly, in the fact that it treats mathematical theories as a result of unification or complication of one or more generating structures. For example, the well-known theory of real numbers manifests itself in such an approach as a complex or multiple structure consisting of algebraic, topological and ordinal structures. Secondly, since the movement from the general to specific mathematical knowledge is, essentially, the movement from individual abstractions to a system of such abstractions, then in this case we also have an integrative process.

Nevertheless, the theory of abstract structure, as the Bourbakians themselves admit, can provide neither a complete nor conclusive idea of the entire wealth of relationships among mathematical concepts and theories because the process itself is to a great éxtent schematic, idealized and static. To comprehend the integrative process in the development of mathematics correctly, one should admit from the outset that abstract structures are changing both in number and, especially in content. Advancements in mathematics result in the appearence of new structures capable of revealing deeper and more significant relationships between mathematical objects, and this reduces the former relationships to specific cases or modifications of a new, more general structure.

In the past few years, the algebraic theory of categories has been attracting more and more attention of mathematicians all over the world. American mathematician S. MacLane, one of the theory's creators, made an attempt not only to generalize, on the basis of the theory, the theoretico-structural approach of the Bourbakians, but also to bring out a new strategy for the basing of mathematics, a strategy which differs in many respects from the established set-theoretical conception. Before MacLane's work, the concept of set was considered the most general and initial in mathematics, Nowadays it is possible to define it using the basic concept "category". Consequently, in Bourbaki's terms mathematics is a "congestion of abstract forms or mathematical structures",[4] whereas with the new approach the concept "structure" proved to have been overlapped, as J. Diedonné, one of the active members of the Bourbaki's school, admits, by such concepts as "category" and "functor" that have upgraded it in a more general and convenient form."[5]

The theory of categories makes it possible, therefore, to treat a direct subject of present-day mathematics as the study of different categories, e.g. groups, field, topological spaces. Even the term "category" itself, as MacLane notes, has been chosen to stress the most general character of the concept denoting this abstract form. To define category it is necessary to specify first a class of *objects* of a certain type. Such objects, for example, can be algebraic groups, topological spaces or even sets. Secondly, these classes of objects should always be specified together with the corresponding classes of morphisms. Thus, when objects of a category are groups, homomorphisms stand for morphisms, continuous mapping are morphisms for topological spaces, whereas all the mapping between them are morphisms for the category of sets. Thirdly, operations on morphisms should satisfy the axioms of association and identity.[6]

A link between the theory of categories and Bourbaki's concept of abstract structures manifests itself in the fact that here we advance to a new stage of abstraction. Indeed, objects of categories are not elements, groups, spaces, sets, etc. of concrete structures, but classes as a whole of such structures. Owing to this it becomes possible to investigate in-depth relations between mathematical theories and, hence, to understand much better integrative processes in mathematics. Such processes can be examined from two viewpoints: internal and external. Integration within mathematics leads to the appearence of new, more comprehensive and general theories, sections of mathematics and even separate disciplines. External integration is associated with the use of abstract mathematical structures to study the real world, i.e. it leads to the integration of abstract mathematical knowledge with concrete scientific engineering or social knowledge.

12.2 INTEGRATIVE PROCESSES WITHIN MATHEMATICS

Advances in mathematics as in any other science are accompanied by its intensive differentiation that leads to the appearance and isolation of new theories and sections in mathematics. By the time mathematics appeared as a theoretical science two independent disciplines, arithmetics and geometry, took shape, the former dealt with the discrete, the latter with the continuous. The inception of algebra and, especially, mathematical analysis resulted in both the immense expansion of the applicability of mathematical methods and the further differentiation of mathematical knowledge. In the late 20th century, differentiation in mathematics intensified to such a degree that many scientists became worried at the possibility of mathematics disintegrating as a single science into several independent theories, division and disciplines with different methods and objectives.

Today specialization in mathematics has reached such a degree that no one can work creatively in several different branches simultaneously; moreover, specialists in a particular field often neither understand the results attained nor even the language used in other fields. Such specialization makes specialists ponder over development prospects for mathematics. If today one group of mathematicians fails to understand the language used and, what is more, the results attained by another group of mathematicians, then we ought to ask ourselves whether it amy not lead in the long run to a kind of tower of Babel and the disintegration of mathematics as a science.

Fortunately, such gloomy forecasts cannot be regarded as sufficiently grounded. In such forecasts people, generally, pay no attention to another important trend in the development of scientific knowledge, the trend usually called integration. Integration within mathematics is always accompanied by the opposite process of differentiation. Whereas differentiation results in the appearance of relatively independent, isolated theories, sections and disciplines, integration, on the contrary, facilitates the advent of integrative conceptions, theories and disciplines.

Integrative processes begin, as a rule, with the development of a common technique or method that can be used as a tool in establishing relationships between those theories and sections of mathematics which formerly appeared as markedly dissimilar in both the objective and the specific methods of research they used.

As a result, new theories or disciplines come into being to deal with problems that were previously considered insoluble. The appearance in the 17th century of analytical geometry, that is the synthesis of ideas of geometries and algebra, is an example of this phenomenon. The integrative mathematical discipline owes its appearence to the coordinate method introduced by R. Descartes and P. Fermat, which made it possible to study geometrical properties of figures using algebraic methods. The same can be said about differential geometry, which appeared later, the only difference being the use of special methods of mathematical analysis, namely, differential calculus to study geometrical figures.

Applying new methods, integrative by their nature, mathematicians could solve many geometrical problems irrespective of their degree of generality. Formerly, to attain this goal it was necessary to look for special methods relying mainly on intuition, at a later stage mathematicians started to use for the same purpose the more effective analytical methods. This dialectical interaction between geometric and analytical methods proved very fruitful for the development of geometry, albegra and analysis. A known example illustrating the validity of geometrical ideas for analysis is the geometrical interpretation of complex numbers that gave great impetus to complex analysis research.

At present the most impressive example of ideas integration leading to the emergence of a new integrative branch of mathematics is the development of functional analysis. It was a synthesis of ideas of algebra, geometry, mathematical analysis and topology that played the leading role in its development. In retrospect, functional analysis itself is the further generalization and development of ideas inherent in an analysis of the *infinitely* smalls proposed by Newton and Leibniz. If, consequently, the 17th century was, according to the apt definition of V. Voltera, the outstanding Italian mathematician, the age of analysis of the *infinitely* smalls, the 19th century — the age of the theory of functions, then the 20th century can be called, not without reason, the age of functional analysis.[7]

All the examples from the past and current developments in mathematics, to mention only a few, are convincing evidence that integration and its counterprocess, differentiation, are crucial factors in the advancement of mathematical knowledge. Many outstanding mathematicians today emphasize this fact with increasing persistence. J. Dieudonné noted that "advances in mathematics owe a great deal to the creative merging of two or more fields of mathematics that are different at first sight. This 'merging' of mathematical disciplines often proceeds evolutionally, but it did take place both in the present and past centuries,... starting from the mid-19th century this process accelerated considerably and acquired new forms".[8] In the development and integration of mathematical theories several common ideas or methods appear, turning, in Diedonné's words, into "radiation centres" as they become applicable to a wide range of different fields.

The way to the common ideas underlying the "radiation centres" was thorny and difficult. Such ideas appeared first as relatively task-oriented and specific because they were interpreted in terms of specific tasks and problems; but step by step their common traits become evident. When a new common-theory appears, such ideas purged of complicated details prove simple and natural. Their simplicity is due to successive abstracting, a process typical of mathematical knowledge today.

Hence, integration in mathematics involves not only an elucidation of relationships between theories and individual mathematical disciplines, but also with the appearance of the more general concepts and methods of research within a new integrative theory or discipline. As pointed out above, synthesis and integration are possible only if there is a common idea, technique or method integrating previously unconnected theories into a new unity and providing the basis for a new integrative theory. To "spot" a new idea or method it is necessary, above all, to establish a similarity of analogy between the theories that so far seemed unconnected with one another. Such analogies and intuitive surmises are the most valuable components in mathematic research. Therefore, when speaking about the abstraction or isolation of common ideas out of a complex context, we should always keep in mind that in a real research process we have an intermingling and interaction of logical, purely mathematical, heuristic and methodological factors. A simple and clear picture of how a new integrative theory comes into being can only be obtained by simplifying and generalizing the real history of mathematical knowledge. For logical-methodological analysis it is important, nevertheless, to focus on the basis, crucial factors of this process, especially the formation of new concepts by abstraction of varying level and generality.

By disregarding the concrete, qualitative content of objects and processing it becomes possible to obtain elementary, initial concepts of mathematics, e.g. natural numbers, elementary geometrical figures. Further advances in mathematics are generally associated with the use of abstractions from abstractions as a result of which new concepts appear by generalizing former concepts. This is how irrational, imaginary numbers, multidimensional and infinite abstract spaces and like objects came into being.

The deeper mathematical research, the more abstract and general become its concepts and theories. This, at first sight, should move mathematics away from reality, but is is well known that the latest abstractions and theories in mathematics proved better adapted to study of most in-depth and ingenious problems facing the natural and social sciences and the humanities.

Why has this happened? The integrative processes associated with synthesis and generalization of concepts and theories have made it possible, in general terms to give, the correct answer to this question. Having established links between different concepts and theories of a less general nature, scientists later single out those important properties and relations which characterize concepts and theories in a more thorough and accurate way than those used in special theories. This means that a theory relying on more abstract concepts proves more adequate to reality. It is because of this integrative ability, which accumulates the potentialities of specific theories, that a new more general theory is more instrumental in solving theoretical and applied tasks.

In this connection a question arises, what are the ways leading to integrative ideas and principles of mathematics since they are very important for both basing and developing mathematical knowledge as a whole? Such integrative force is sometimes associated with the deductive method of mathematical proofs. It is but natural that deductive inferences distinguish mathematics from the natural and other exact sciences generally using analogies, comparisions and deductive generalization, which per se have no demonstrative power. It is evident, however, that the deductive method of drawing conclusions on the basis of previous inferences makes it possible only to arrange the available knowledge in logical order, but it has nothing to do with the general premises from which this knowledge follows. As N. Bourbaki aptly noted, deduction "cannot, of course, be an explanation of unity of different mathematical theories, just as it is impossible, for example, to unify physics and biology into one science on the ground that they use experimental methods".[9]

Hence, a unifying principle should be looked for in the ideas that enable a researcher to deduce available knowledge in a particular mathematical theory with the help of deduction rules. This is why deduction plays a secondary role in the origin and integration of mathematical knowledge. The discovery of basic ideas and principles that become prerequisites for further deductions is the most difficult and truly creative component of mathematical studies. As in any study, the researcher's intuition and experience, heuristic and inductive guiding considerations are of prime importance here, though in the established theory all references to such considerations are believed to be not only unnecessary but even inadmissible. Since mathematical theories are generally expounded in axiomatic form, their basic principles are given in the form of axioms or postulates. Bourbaki even believes that it is the axiomatic method that teaches us how "to find common ideas which remained undiscovered due to details inherent in each of the theories in question, to expose these ideas and to study them".[10] Though the axiomatic method as such cannot be a means of discovering new scientific truths, nevertheless, it plays an important role in the systematization, streemlining and logical organization of scientific knowledge. It would not be out of place to mention here that the concept of abstract structures itself relies on the extensive use of the axiomatic method. Its integrative role manifests itself, first, in the unification of previously isolated and unsystematized knowledge within a common system built up with the help of relations of logical deduction; secondly, in the abstract and formal treatment of axioms they become forms permitting a wide range of different interpretations and, hence, capable of describing from a common viewpoint a great variety of real processes; thirdly, using abstract forms relying on the axiomatic method, it is possible to relate the abstract to the concrete and the formal to applied knowledge. Thus, the integrating role of abstract structures manifests itself in full when mathematical methods are applied to the study of properties and relations in the real world, though this generally takes place when mathematics is used in the natural sciences and in the past decades in the social sciences and humanities.

12.3 INTEGRATIVE PROCESSES AND MATHEMATIZATION OF SCIENTIFIC KNOWLEDGE

We have been discussing so far problems associated with integration within mathematics itself. Henceforth attention will be focussed on how such internal integrative processes effect the integration of scientific knowledge as a whole. Their effect takes place, of course, when mathematics is applied to a particular science. This is why such aspects of integration may be called internal integration.

First of all, it is necessary to note that a possibility of applying mathematical methods to other fields of knowledge presupposes the integration of mathematical concepts with concepts used in a particular science. Indeed, mathematical formulas, equations and structures take no account of the concrete qualitative nature of the real world. Consequently, taken in "pure form" they can neither be verified nor used for the study of the real world. Since the qualitative and quantitative, applied and formal, single and general properties and relations of the real world exist in their intrinsic indissoluble unity, we shall be able to adequately understand real processes in the world only by taking into account the dialectical relationships of both mathematical and specific scientific methods of research. This is why the use of mathematics requires that its abstractions should "be filled" with real content. In a real-life situation, this general requirement is met while interpreting variables of equations or more complex mathematical structures.

The integrative nature of processes associated with the use of mathematical methods is evident in mathematical modelling. Such models belong to conceptual, sign systems reflecting certain significant properties and relations of real processes and, therefore, they can, to a certain extent, replace them, facilitating research considerably. Instead of dealing with a physical or even "quantitative" conceptual model, the mathematical model makes it possible to scan many different versions to select the optimum. To this end the mathematical model with specified parameters is processed by a computer and the results obtained are compared with one another and then with real-life requirements. This research technique is generally called a mathematical, computer-based or machine experiment because it is partly similar to the real-life experiment, especially to the imaginary experiment. Mathematical as well as imaginary experiments were used in the past too, but they received new dimensions only with the advent of high-speed computer technology since computation of complex mathematical models requires a great number of calculations. The mathematical experiment regarded in the light of the paper' topic is one of the most efficient integrative methods of theoretical and partly practical research using various abstract structures implemented in corresponding models.

In building a complex model of real processes in science and technology, not to mention integrated or global modelling, it is necessary to analyze many variables, to single out most efficient ones and to offer an equation of mathematical relationships among the variables. There will be no success, of course, without the concerted effort of a subject specialist or even a team of specialists and knowledgeable mathematicians. This means that the efficient use of mathematical models is possible only when mathematics has been integrated with other sciences, and today with computer technology, too. The optimum here is, of course, the situation when mathematicians are familiar with a given problem area, while the experts directly involved in the research understand how to use mathematical methods. Apart from this, integration means not only a concerted effort but tasksharing in achieving a common goal as well. In mathematical modelling of a phenomenon, experts are concerned mainly with the elucidation of variables characterizing the phenomenon, setting apart basic and significant variables and establishing quantitative relationships among them. The mathematician's task here is to find the exact quantitative or structural relationships between the variables. To this end, either the appropriate equation or the more complex abstract structure, if non-metric relations are analyzed, should be found. Nevertheless, since neither equation nor the more complex abstract structure describing a phenomenon mathematically permits their empirical verification, the mathematician should deduce several special corollaries some of which allow experimental verification, and this is the task of an expert.

Let us now consider how concepts and methods used in the theory of structures facilitate the application of mathematics to other sciences. As shown above, such structures take account of neither the real-life nature of subjects or processes in question nor relations among their components. It means that such structures can be used for the mathematical analysis of a wide range of different objects. To do this it is necessary only to interpret as required concepts and axioms of the structure. With such structures at hand, the researcher's task boils down to verification of whether or not the phenomenon under study meets the structure's axioms. For example, revealing phenomena possessing the property of symmetry, the researcher can be sure that a group-theoretical structure can be applied to them. Any further labour consuming work to deduce the axiom's corollaries turns out to be unnecessary since even a non-mathematician would be able to use at once all the theorems obtained in the mathematical study of the structure. Thus, abstract structures can be compared to ready-made forms, which can be used in studying a wide range of different objects and processes. This widens and facilitates considerably the use of mathematical methods.

Another approach is associated with the use of abstract structures for mathematical modelling, especially modelling aimed at the elucidation of both quantitative (metric) relationships between variables and relationships of a non-metric character. The study of non-metric relations is of paramount importance for the sciences whose results cannot be quantified either because the object under study is too complex or because a developed theory in the field is not yet available. In such a case researchers have to limit themselves to the use of order structures, including comparative concepts, i.e. when such terms as "more", "less" or "equal" are used to show relationships.

Thus, psychologists and sociologists are now using the graph-oriented theory, that is an elementary type of the algebraic category to study various relations of subordination and hierarchy in social groups.

The most important developments in the mathematization of such sciences as biology, psychology, economics and sociology can be expected due to the use of non-metric mathematical models based on the concepts and methods of abstract structures. Many biologists nowaday are beginning to be aware of this. E. Moore noted that the "special usefulness of mathematics for biology consists... in the possibility to abstractly approach the solution of basic problems and to elucidate relationships between principally different phenomena and processes".[11]

Experience in using structural methods in exact sciences convincing evidence of good prospects for this trend in science mathematization. Indeed, the use of abstract structures in such fields of natural sciences as the theory of relatively and quantum mechanics, the elementary particles theory and cosmogony, quantum chemistry and molecular biology, and some other sciences, is dictated by the development level of the sciences. The concepts and theories they use generally allow neither graphic representation nor their treatment in terms of classical mathematics. Thus, the example of abstract structures has proved the apt idea of V. I. Lenin that scientific abstractions do not get us away from the truth but bring us closer to it.

The integrating role of present-day mathematics in general and of abstract structures in particular is most evident in establishing integrative disciplines in the field of engineering knowledge. The history of cybernetics as a science of common principles of control is convincing prove of this process. To reveal common principles one should abstract from concrete, specific features in a real-life control system. This, in turn, requires the use of efficient mathematical methods. We know that no breakthroughs were possible in cybernetics and advanced control methods to manage technological and other processes in industry and social life, no automation of certain intellectual processes could be attained without the quantitative theory of information, the theory of algorithms, mathematical logic, and the theory of models. The key integrating role in this development belongs, of course, to the information theory which has made it possible to describe mathematically a wide range of processes associated with storing, obtaining, processing and transmitting information. A. N. Kolmogorov, Soviet Academician, emphasizes that "cybernetics deals with the study of any systems capable of accepting, storing and processing information to use it for control and regulation".[12]

The development of a common information language theory has made it possible to reveal important features common to all control systems without exception. Likewise, the accurate definition of algorithm as a mathematical concept has made it possible to unambiguously describe various procedures and modes of operation generally used in research and practical activity. Cybernetic control systems are important though not the only systems coming within science interest. Applied systems analysis stemming from operations research also relies to a great extent on the latest mathematical methods, especially in studying modelling and decision-making. The difficulties the general theory of systems is facing can be attributed mainly to the absence of a sufficiently general and adequate mathematical apparatus to analyze regularities in devising and operating a large number of standard systems in science and industry. It is because of these inadequacies that theory is generally critisized. As far as applied systems analysis is concerned, which is used extensive both in designing complex engineering systems and numerous task-oriented systems in science, their adequate description is assured by the use of present-day mathematical methods. Naturally, such a description often runs into difficulties, of which most important is the absence of a logical-mathematical apparatus to expound certain common features of complex dynamic systems, e.g. living and social systems. That is why in studying such systems scientists in certain cases have to describe processes in a sketchy and schematic way.

Despite these difficulties in the mathematization of science, the concrete integrating role of mathematics can hardly be denied. This is evident from the increasingly wide use of mathematical methods and computer technology to solve topical multidisciplinary problems of science and technology, ranging from organizational aspects of production management to such global problems as the efficient use of natural resources and environment protection.

REFERENCES

[1] D. Hilbert, *Grundlagen der Geometrie*, Moscow-Leningrad, Gostekhizdat, 1948, p. 55 (In Russian).

[2] N. Bourbaki, *Essay on the History of Mathematics*, Moscow, Mir Publishers, 1963, p. 251 (In Russian).

[3] Ibid., p. 255.

[4] Ibid., p. 258.

[5] J. Dieudonné, "On Bourbaki's Work", in: *Advances in Mathematics*, vol. 38, issue 3, p. 209 (In Russian).

[6] P. Kon, *Universal Algebra*, Moscow, 1968, p. 49—50 (In Russian).

[7] *Problems in Mathematics Today*, Moscow, Znanie Publishers, 1980, pp. 37–38 (In Russian).

[8] Ibid., pp. 5–6.

[9] N. Bourbaki, *Essay on the History of Mathematics*, p. 247.

[10] Ibid., p. 248.

[11] E. Moore, "Mathematics in Biological Research", in: *Mathematics Today*, Moscow, Mir Publishers, 1967, p. 130 (In Russian).

[12] A. Kolmogorov, Foreword to Russian edition of *Introduction to Cybernetics* by W. R. Ashby, Moscow, 1959, p. 8 (In Russian).

13. Systems Theory and Cybernetics

V. S. TYUKHTIN

The appearance of the systems theory and cybernetics arose from the necessity to solve problems of integration of scientific knowledge and realization of interdisciplinary investigations. Much has been said about it by the pioneers of systems and cybernetics investigations (A. A. Bogdanov, L. Bertalanffy, N. Wiener). The chief thought of interdisciplinarity determines investigations of systems theory and cybernetics, their horizons of development. Systems and cybernetics investigations influence one another because of their common ground — the systems approach. Interconnections between the systems theory and cybernetics, as far as we know, were not relevantly considered in modern methodological literature. This problem is the subject of the present chapter.

13.1 SCIENTIFIC STATUS OF SYSTEMS APPROACH AND GENERAL SYSTEMS THEORY

Any empirical object, reflected by the philosophic category of the particular, is an organic unity of the general (of different levels) and the singular. The general expresses primarily that every object belongs to different classes or systems of objects of the same kind and, secondly, that any object, considered by the knowing subject, is polysystematic, i.e. it may be decomposed in different systems with regard to the nature of the object itself and the aims of the subject. These different systems as images or "projections" of the same object constitute the field of investigations of different sciences and scientific disciplines. For example, the cell is the object of analysis in biophysics, biochemistry and also of morphology, physiology and other biological disciplines.

The object of knowledge is also polysystematic in another sense: the plurality of systems, that may be found in the contents of this object, depends upon what we, according to our aims of knowledge, take as basic (non-divisible) components of the system and, accordingly, elements of the structure.

The choice of the components is connected with the choice of a system, both are determined, we would like to emphasize it once again, not only by the nature of the object, but also by the aim itself. In dependence with the character of the systematic purpose, we shall include the object in question in the more broad systems of different levels of generality (the extrospective systems analysis). The set aim determines also the depth of the inner division of the object in systems and subsystems of different levels, that ultimately helps to find the indivisible (further non-divisible) component of the system (the introspective systems analysis). For example, the task of the physiological explanation of the reaction of the brain is connected with the analysis of interdependence of neurons, which are the components of the system. The task of explanation of the substratum stability of the neurons leads to the levels of molecular physics, biochemistry and even quantum chemistry. In other words, the notion of the system, structure and component (element of structure) has two aspects of analysis, that is ontological and logico-epistemological. The break between them leads to subjectivism or descriptivism, irrelevant to the aim set, that is without any heuristic value. Thus the notions of the system and structure, component and element, reflecting reality, are at the same time abstractions, connected with the aims of the subject.

Interconnections of objects (components) under certain circumstances serve to realise integral qualities of the system as a whole, different from particular qualities of its components. That means that the systems approach is directed differently than the aggregational approach, for which elements and their qualities are primarily set; thus certain characteristics of the aggregate are identical with the qualities of the elements, and the other are equal to the sum of the characteristics of the parts (the addition of qualities). A system, which has a smaller or more large degree of integrity may be acknowledged, starting from the functions of the whole, upon which qualities and quantity of elements depend and, therefore, the division of the system in its parts. But the difference of the aggregational and the systems approaches is not absolute in the sense that the influence of the elements on the whole and all the system as a whole on its parts is mutual: the whole is the effect of interdependence, interconnections of its parts; parts are influenced by the whole through immediate and mediated interconnections between them. Therefore the acknowledgement of the leading roles of parts (elements) or of a system as a whole are complementary poles of the unitary systems method. For simple, in particular, linear systems, in which mutual influence of the parts is not substantial and their relative independence is more evident, the first approach prevails; for systems, in which mutual influence, interdependence of elements and subsystems are great, belonging to different types of nonlinear systems, the systems approach prevails in the proper (narrow) sense. And as far as the character of interdependences and interconnections of elements, the whole on elements and elements on the whole is expressed by some "kind of arrangement", "kind of composition" of elements, stable about certain variations (changes), that means that the key notion of the systems approach is the notion of structure. Some authors even call this approach systems — structural, including in it the stage of the structural analysis.

Mutual addition of the roles of the elements and the whole demands the description of the real systems a unified language and in a certain sense, a language of compromise. This language is a language of mathematics, in the foundation of which the theory of multitudes lies. Though elements of the multitude are given in abstraction initially and independently of the integral properties (characteristics) of the multitude, mathematical methods (algebraic, that of mathematical analysis, numeral methods, etc.) give the opportunity of adequate description of mutual influence of parts on one another and on the whole. These are different in their form non-linear dependences: nonlinear algebraic, differential, integral, differential-integral equations and relevant to them diagrams and tables. Another thing is that for the description of organic-whole systems a special language is needed more immediately expressing the idea of the primacy of the whole to its parts. With the help of such a language problems concerning functioning and development of organic-whole systems may be laconicaly described and effectively solved. The present stage of mathematical description of organic-whole systems may be compared with the attempts to describe the movement of the planets around the sun on the base of geocentrical conceptions. Unwieldy apicycles are "additional difficulties", connected with the inadequacy of the language to the character of the described objects.

However, in the language of modern mathematics organic-whole objects are described and specific systems problems of two types are solved.

(1) To define integral properties of the systematic whole, knowing the interconnections of elements (and subsystems) and their properties. In other words, the integral properties are mathematically expressed as functions of properties of elements and structures (laws), their interconnections.

(2) To define the properties of elements, knowing the properties of the whole and structures (laws) interrelations of elements and subsystems. Such systems' qualities of elements were called by K. Marx systems qualities (see 1).

If the main task consists in the investigations, definition and usage of structures of system-object, we can define the essence of systems-structural approach, which will include the following three main thesises.

1. The knowable, transformed or projected object in principle may be systematically (polysystematically) interpreted, that is considered as a system (or a multitude of systems) or as an element of a system, in which it is included (for example, such are the elementary particles or hypothetical quarcs).

2. According to the nature of a system-object as simple or complex, aim-oriented or non-aim-oriented, any other of its specific properties, we must work out a strategy of its knowledge, transformation or projection and apply appropriate methods.

3. Any properties of an integral system depend on the structure of different levels (or facets) of organization of its elements. In its turn, properties of components of the system depend on the structures of the system as a whole. In both cases integral properties of the system as a whole and systems properties of particular elements and subsystems depend on relevant structures. That is why the knowledge of structures and structural analysis of properties of the parts and the whole are preliminary conditions and a certain link (stage) of the mathematization of knowledge.

The realization of the systems approach includes in itself gradual solvation of the chain of problems, the most characteristic of which may be summarized as follows: — the definition of the boundary of the system and its surroundings (the choice of the upper boundary); — the definition of the limit of the decomposition of the system (the choice of the lower boundary of the division into components); — the definition of the basic properties of the components and subsystems of the system and their aims or functions within the whole system (tables are done for subsystems or a finite number of elements, revealing their connections and functions in the integrity of the system); — considering the outer and inner conditions (aims in particular) of the system, its structures are defined; — on the basis of knowledge of the structures all the properties of the system and elements are expressed as functions of these structures; — the definition of the programs of its possible behaviour ("matrix of behaviour").

The systems approach to the solving of such problems in the investigation of some system-objects needs the strict language, in which the conceptual and mathematical (mathematical methods) set, systems set, represented by definite statements, theorems, formulas, characterizing systems are included. Such instruments and the language are the products of the general systems theory. But at present there is no universally accepted general systems theory (GST) and there exist some variants, having their own conceptual and logico-mathematical set. The plurality of variants is the result of differences of approaches of its authors to the scientific status and the choice of foundations of the theory (2).

The main difficulty in the building of GST, differently orienting its creators, consists in the understanding of the general character and the universality of the systems approach and the general systems theory. If the general character refers to the sum of systems notions, that is the range of known objective fields of reality, its universality is related to the contents of the notions.

The universality of the contents of the systems notions follows from the essence of the ideal of the systems approach — firstly, from the possibility of systems interpretation of any object, secondly, from the existence of correlation between any characteristics, properties, from one point of view, and structures of different levels or "cuts" of the organization of the object — from another. In other words, with the help of structures, as shown further, the inner contents and functioning, integral properties and functions in the outlets of the system, its exterior behaviour, genesis and evolution can be expressed in principle. The universality of systems structural analysis and the notions of the system and the structure is immediately connected with the universality, range of variety of understanding of the notion "component of the system" and "element of the structure". In the roles of the components of the system considered by us may be things (i.e. material particles, units of this object), properties and states, connections and relations, phases and stages of the inner functioning or the outer behaviour of the object, cycles and periods, levels of development of the system-object. With dependence on the choice of the type of these components of the system, we deal with such systems and their relevant structures as substratum (statical), parametrical, relational (for example, of space and time, cinematic), functional, behavioural and, finally, with systems and structures of the processes of genesis and development of the objects.

The narrowing of the understanding of the notions "system's component" and "structure's element" by some authors considerably limits the sphere of application and contents possibilities of the systems structural analysis. For example, the opinion is widespread, that the systems analysis may be applied only to the questions of structure and functioning, but not to the development of systems.

With the understanding of the universality of the systems approach and the notions of the general systems theory is connected with another prejudice — the fear of the confusion of philosophic categories and systems notions, philosophic and systems approaches and analysis. But they are radically different. The method of philosophic analysis is characterized by all-round approach: the system of philosophic categories reveals all the richness of qualitative multiplicity of the world, universal forms of being. The systems-structural approach and analysis of object are characterized by a certain one-sidedness: all the spectrum of the qualitatively different properties, characteristics is expressed only through the one trait of the contents — through the different kinds of organization of systems, through the difference of types of structures of the system-object. A qualitative variety of components and types of relations is expressed in a relatively abstract — symbolic, sign form. Different from the sciences, having their empirical basis, the limited objective field of investigations and, hence, expressing properties in the measurable form, general systems analysis expresses properties in the non-measurable form. Due to that the systems-structural approach and the general systems theory help to reveal strictly the resemblance of laws, isomorphism of structures, referring to different spheres of reality. (For example, isomorphism of formulas of laws of Ohm, heat condictivity of Fourier and for filtering liquids of Darcy.)

Notions and categories of special sciences and philosophy can be interpreted in a systems-structural way, helping to make strict their contents, giving them unique sense (3, 4). For example, structural interpretation of categories of the cause and the consequence helped to find that the thesis that the consequence can produce an effect on the cause, that created it, is not correct. But the consequence can influence the bearer of its cause, not the cause itself. The last thesis is against the postulate of the non-reccurence of time. We have the analogical example with the traditional understanding of form as a structure of the contents. Such an interpretation contradicts the phenomenon of the isomerism which is widespread in the inanimate nature, animate nature and society. The essence of the isomerism lies in the fact that with the same components and the different structure we have qualitatively different objects. And if it is assumed that the form is the structure of the contents, we shall inevitably reject the materialistic thesis of the primacy of the contents over the form. In reality the contents and the form (for example, in art and science) have their own structures; the structure of the form depends on the structure of the reflected contents.

The presence of the resemblance between structures (structural expressions of laws in particular) of different objective spheres of reality, principle possibility of the structural-mathematical expression of any property and also systems interpretation of any scientific notions and philosophic categories — all that shows the material unity of the world, expressed in the philosophical-methodological principle of the systems organization and the systems approach as a special methodological and interscientific principle. These manifestations of systems organization are the foundations of the thesis that the systems principles, notions and logico-mathematical set serve as an interdisciplinary, special scientific language.

With the signified universality of the systems approach and GST another characteristic is connected — generality of systems notions and methods, i.e. their applicability in all spheres of scientific knowledge.

The differences in the interpretation of the generality of notions influenced the ways to the creating of the general systems theory. The most urgent and vulnerable question is about the connection of the generality of GST with its effectiveness, that is capability of forcasting function, ability to find strict programs of human actions. Many specialists in the systems theory say that the generality (uttermost universality) and extreme abstractiveness of the notions and theories, corresponding to it, led to trivial scientific results.

One of the authors of the variant of the GST, M. Mesarović, writes: "It is evident, that the more abstract a statement is the more wide the range of objects it can cover, but, at the same time, the less information it carries about the behaviour of any concrete system. Thus the main difficulty in the creation of any general theory consists in the choice of the needed level of generalization or abstraction" (5, p. 18). Analogically K. Boulding says: "We always sacrifice the contents to the generality... But between the specific, with no meaning, and the generality, with no contents, there shall exist independently of concrete aims and of the degree of abstractiveness an optimum level of generality" (6, p. 107).

According to G. Simon the emptyness of the notion of a system "lies in the fixation of the useless because of the excess degree of isomorphism" (7). "We have not reached yet, — says E. A. Feigenbaum — the satisfactory degree of understanding of the problem of the generality of the set aims. The analysis of existing programs of solving problems, as well as the common sense, show that here a certain kind of the "law of nature" works, linking with inverse proportion, the degree of generality of the method of solving (the size of application) and its power (the possibility of reaching results, effectiveness etc.), and, with direct proportion, — the power of the method and its specialization (information, expressing the specific problem)" (8, p. 175).

It is evident that in these statements such general notions are meant, that are based on Locke's abstraction of identity, that is in these notions signs identical for the class of the objects of generalization are fixed but signs specific for each class are excluded. In other words, generalization of such abstract-general (differential-general) notions consists in neglecting specific signs and the limitation of the generality of notions — in the addition of the new signs. For this kind of understanding of the generalization and limitation of notions in the form of logical subordination relations (genus-kind relations) the inverse relation between the contents and the range of the notion is characteristic, as the cited authors underline.

Nevertheless the history of the scientific knowledge, especially of the formation of fundamental theories, presents evidences of the existence of the other type of generalization and formation of general notions and methods. These are concrete-general (or integral-general) notions and methods. For them not only the relations of subordination or genus-kind relations (vertical) are characteristic, but also relations of subordination, when the given general notion "is served" not by one, but several less general notions (horizontal subordination). For example, the notion of symmetry has as subordinate notions of the lesser degree of generality three types of symmetry: geometrical, structural and dynamic. Each of these three types of symmetry is divided into two types of classical and non-classical symmetries; they, in their turn, have branches in every subsequent lesser degree of generality. Thus, the notion of symmetry is multi-level in the degree of generality, having the "tree" of many branches (graph) of notions with different signs of symmetry. In short, the general notion of symmetry includes in itself a variation of the different notions of symmetry with their specific signs. That means that in the process of the structural analysis of the system-object it is possible to find specific symmetric signs of the given class of objects on the way of concretisation of the notion of symmetry and to apply appropriate mathematical methods. Let us consider another fundamental integral-general notion from the sphere of theoretical physics — the notion of the smallest effect. From equations of Hamilton, corresponding to the principle of the smallest effect, there are inferred, by the way of putting empirically suggested limits, all the main structures (laws) of theoretical mechanics.

Specialists, treating general notions as abstract-general, suggest different ways to avoid non-significant generalizations and trivial results (see : 9). Thus, M. Mesarović, wanting to avoid abstract generalizations of the notions of GST claims that this theory "shall unite special theories, devoted to more concrete class of models, for example, to the theory of linear systems, theory of Markov's systems etc." (5, p. 20). "The general systems theory, — he continues, — unites also theories of different aspects of behaviour of the systems, such as the theory of relations, the theory of control, the theory of adaptation, of selforganization and learning, the theory of algorhythms etc." (5, p. 39). But GST, which on its scientific status is based on collection and generalization, is limited in its heuristic possibilities in comparison with the theory, based on the fundamental integral-general notions.

Another way of limiting of the generality of notions of the systems theory, according to some authors, is possible to obtain by identifying it with the theoretical cybernetics (10). Close to this point of view is the limitation of the sphere of the systems theory by the field of complex and organic whole systems.

The third way is connected with the shift of attention to metatheoretical and epistemological aspects (11, 12). Here the general systems theory is in the role of "the general theory of systems theories" (12, p. 84).

There are some other ways of limiting of the generality of the systems theory in the ontological aspect. For example, the purely empirical way of generalization without addition of deductive components that cannot pretend to be strict and complete GST. Thus, the variant of GST of L. Bertalanffy has, according to his words, "empirical-intuitive character" (13, p. 33). He formulated some empirically evident isomorphisms in the structure and functioning of different system — objects, but that could not lead to the creation of the formal mathematical set on the level of generalization.

In literature devoted to systems investigation there are two variants of the general systems theory, which realize the unity of the generality and effectiveness on the basis of the specific conceptual, logico-mathematical set and general methodological premises. These are variants of GST created by A. I. Uyemov and U. A. Urmantzev, which are regretably poorly known by the specialists in the West because of the language barrier.

In the variant of GST worked out by A. I. Uyemov the main accent in the description and analysis of system-objects is put on the revealing of general sign-systems parameters, their classification and definition of the so-called systems laws (14). The latter are correlations of two, three, four or more systems parameters (for example, such as stability, centering, the chain character of systems, regeneration in relations, etc.). Systems laws are fixed with the help of statistical methods, applied to empirical facts about different systems with the use of the computers. In their character (type) generalizations of Uyemov in his variant of GST are abstract-general theory: it describes different system-objects not through specific parameters and laws (structures) of this class of objects, but by means of complexes of systems laws of a high level of abstraction. Such description sets aim at the choice of specific methods and mathematical language.

U. A. Urmantzev's variant of GST has other foundations and scientific status. This theory is concrete-general (or synthetic-general) and is effective from the mathematical point of view. General and at the same time specific structural characteristics of the systems in the variant of GST of Urmantzev are polymorphisme (including isomerisme) and isomorphisme, symmetry and assymetry. He notes the laws of polymorphization, isomerization, correspondence and symmetrization (15, § 3, 4, 5). From the one side, these are mathematical notions and structures, having integral-general character, that is they are graphs of the notions, representations of the gradation of signs from the universal level to the levels of different degrees of generality and concreteness. From the other side, these notions and structures are interpreted in the systems theory as fundamental, attributive properties of matter, having the empirical basis. For example, symmetry-asymmetry properties of biological objects in the recent 15—20 years received their systematic analysis in the biosymmetry — fresh scientific discipline. Basing his investigations on the correspondence of the symmetry-asymmetry signs with the empirical signs of the animate systems and on the laws of polymorphization, isomerization and symmetrization, Urmanzev predicted new empirical signs, behaviour or biosystems and the change (control) under different conditions (see 16).

In spite of the difference of approaches to the creation of GST and between already made variants of GST it must be noted that the universality, generality and the different degrees of effectivity characterize general systems theories from the point of view of theoretical functions — calculating, projecting, explanational and integrational-systematizing. These functions are expressed in different degrees in these variants in dependence with the initial basis. At present we witness the rapid growth of the conceptual and logico-mathematical sets in different variants of GST.

Systems theories have not only functions, but also the special-methodological function in connection with the variety of scientific disciplines. Any variant of GST is dependent on the definite interpretation of the systems approach, fused with the applications of GST to solving concrete (theoretical and practical) problems, with the complex of methods in the form of the so-called systems analysis. The unity of the systems approach with the conceptual and mathematical sets of the systems theory and based on it methods of systems analysis forms systems method with its special-methodological and theoretical functions.

Because of the generality and the universality of the systems approach, notions and methods of different variants of GST, the systems method is in immediate contact with the scientific philosophy — with the philosophical laws and categories of the dialectical materialism, with the epistemology and its parts — the theory of reflection and the methodology of science. Notions and methods of GST are also in immediate contact with the categories and principles of the dialectics, being their "twins" to some extent. For example, the philosophic systems principle corresponds to the systems approach, to the principle of the preservation and change — the principle of symmetry, to the law of the unity and struggle of the opposites — the chain of twin categories (element — system, part — whole, structure — function, system — chaos, polymorphism — isomorphism, symmetry — asymmetry, etc.). To the philosophic categories of the system, the structure, the element, the part categories of GST correspond with the same names.

The analysis of relations between philosophic categories and corresponding to them systems notions that are in contact with them is very important. These systems notions must reach maturity, the level of integral (synthetic) generality; because of their many levels ("the tree of dependent and interdependent notions") they are in contact with the special scientific notions, influencing them methodologically, defining methods and operations with them. In another words, systems method is a transitional link, a conductor of the general methodological strategy, a transformation of it in the tactics of knowledge, projection, prognosis, planning and control. That means that the systems approach and GST, in their turn, need a mediator, which can produce concrete formalized system schemes, models, formulas and apply them to the analysis of different spheres of nature and society. To such scientific trends and disciplines belong cybernetics, the theory of communication and the theory of information, the theory of identity and size, informatics and the science theory and also a number of new parts of the modern mathematics (theories of algorithms, models, structures, categories and functors, topos, etc.).

13.2 THE RELATION OF THE SYSTEMS THEORY AND CYBERNETICS. TENDENCIES OF THEIR DEVELOPMENT

The first stage of the systems investigations is connected with the creation of the first variant of the systems theory by A. A. Bogdanov in the three parts of his tectology (17), published in the twenties.

In the thirties (the second stage) the idea of the systems approach and the building of GST was revived in the works of L. Bertalanffy and other scientists (it is regrettable that without mentioning the books of Bogdanov).

The third stage is connected with the birth of cybernetics (N. Wiener's book "Cybernetics" was published in 1948). The scientific and technical designs of controlling device, the experience of working out and exploitation of automatic and automatized systems gave rise and the solid ground for a real renaissance of systems investigations and their full development. On this stage the mutual influence of cybernetics and the systems investigations started, also the mutual enrichment and interpenetration of notions and methods. It was caused by practical problems and tasks. Enough to review the application of the systems designs, methods of systems analysis, so that to find the unity in the complexes of the ideas and methods of cybernetics and systems investigations. It is evident that different divisions of cybernetics influence systems investigations in their own way and themselves, in their turn assimilate the achievements in this branch of knowledge. On the third stage of systems investigations the direct influence on them on the part of cybernetics still prevails, starting from the fusion of cybernetics and the systems theory in the works of O. Lange, the appearance of cybernetic problems in the systems theory (the systems theory, according to W. R. Ashby, is a theory of simplification, because of creation of the theory of big systems in cybernetics) to the inclusion of some notions and methods of cybernetics in the systems theory. To them belong such notions as backward relation, aim (aim function), optimisation and suboptimisation, making of decision, algorithm and program, inlet and outlet of the system, local and spread parameters and fuctions etc. Especially great appeared the mutual influence of the systems technique (the theory of big technical systems) and the systems investigations.

In the sixties and seventies there were created original versions of GST with their own notional and logico-mathematical sets. This was the fourth stage of the systems investigations, of which differentiation and isolation of GST from one another is characteristic, precision and detalisation, widening of the mathematical set for solving of new problems. The process of interaction with cybernetics has become more precise and constructive. Besides, the book on the mathematical foundations of the general systems theory (18), M. Mesarović and his coauthors published the book devoted to the analysis of hierarchical multilevel systems (19), representing in itself a division of theoretical cybernetics.

After a brief review of the main stages of systems investigations and their relations with cybernetics we shall consider the essence of their mutual ties.

Cybernetics studies a special type of systems — systems of control and the control on the basis of the necessary information; to them belong systems of animate nature and society, artificial (technical) systems of communication and control as a special kind of products of society. Unlike classical automatics and telemechanics cybernetics studies mainly complicated and overcomplicated systems. A special class of complicated systems compose the so-called big systems, which are the object of the analysis of special part of cybernetics — systems technology.

Systems technology is close to the general systems theory, in the respect that technical control systems, viewed without man, who creates, uses them in this purposes, repairs and controls, are peculiar abstractions embodied in metal. Without man it is metal, having certain forms.

Still closer to the general systems theory is theoretical cybernetics, which is an abstract systems theory of working over information for the purposes of control. We have already said that the integrative, interdisciplinary role of GST was closely connected with the character of the generality in the range and the universality and the informational value of the notions of GST. It is not enough to characterise cybernetics as a less general and less abstract theory in comparison with the general systems theory. If GST may be regarded (as M. Mesarović did) as a uniting-general theory, then cybernetics is a part of GST, in which such systems as systems of control are studied. If GST is the analytical general theory, considering no specific classes of systems, then cybernetics studies a kind of systems, the knowledge of which is acquired empiricaly.

At last, if created GST, in the sum of its fundamental notions and methods, has grown mature and reached integral-general (synthetic general) stage, then many theses of cybernetics (and also the biology and other disciplines) may be worked out deductively from the notions and laws of GST. For example, from the variant of Urmantzev's GST many theses of Darwin's theory of evolution follow (16).

Analogically, for the growth of the informativity, the integrative — interdisciplinary function of cybernetics, all its notions and methods, belonging to different parts, must become integral-general notions. Let's take an example from the history of the formation of the general theory of recognition of images (classes) in cybernetics.

Very often theses of a purely unitary character have been expressed: the general theory of recognition must unite already existing methods and algorithms, part of which has the intuitive-empirical basis, the basis of the method of tests and errors. Another way of building of the general theory of education and recognition consists in the application of generalisation and synthesis. Soviet scientist J. Z. Zipkin found the integral-general algorithm on the way of the probability iterative approach. From this algorithm, as the author shows, more particular are deductively inferred, including the well-known algorithms of the self-education of recognizing objects (20, p. 128—135).

Nevertheless the great part of notions and methods of cybernetics are still on the level of analytically general notions or notions of cumulative (integrative) generality. That does not reduce the integrative-heuristic function of cybernetics, but leads to errors and incorrect use of analytic-general notions above the boundaries of their applicability.

We may see it in two cases of formation of notions of information and control. The transition from inanimate systems to the animate, aimoriented systems, capable of selfcontrol on the basis of the use of information, is not a simple addition of another new class of system-objects, not a simple acquisition of two signs, limiting the generality of the abstract notion of the "system as such". With the appearance of a new signal-informative factor old properties substantially changed and emerged new properties and functions of the bearers (transformers and generators) of the information: the forestalling reflection, the aimoriented behaviour, the growing of the rate of integrity and, hence, the variability and the dynamic stability, the sphere of the spreading, the tempo of development etc. Let us see the role of the informational factor in the growing and change in the character of the integrity. The integrity of the inanimate systems is an effect of the mutual influence of elements and subsystems, which, in their turn, through such relations are under the influence of the whole system. The integrity of an animated being is embodied first of all in the genetic information, controlling the growth of the embryo and processes of the differentiation, integration and growth of its cells, tissues and organs. The primacy of the whole over the development of its parts is the expression of a new type of the integrity — the so-called organic whole. Further the appearance of the signal-informational relations with the surroundings as a factor of the orientation in the environment, advanced reactions to the future actions and their results became possible on the basis of the highest level and type of organization of the systems — their self-organization. The selforganizing systems are systems, having activity — life and psychics. In other words, the actualisation of the signal — informational factor is principally possible in the active systems or with the help of the active systems. That means that abstraction from the properties of the bearer of information is not absolute, but relative, that is justifiable only within the limits of purely technical tasks.

The presence and work of the active systems in the realization of the informational relations is determined by the fact that the results of the interrelations of the reflecting (informational) system become the reflection in the proper sense only when these results are used by the active system. For all that, firstly, from the imprints, traces, reactions is functionally extracted what is dependent on the reflected objects and the characteristics of the bearer of the reflection are excluded; secondly, the resemblance of the extracted characteristics of the reflected original is actualized and corrected; thirdly, extracted characteristics (of relation and structure of different degrees of modality) are objectivized, related back to the original, that is acquire objective sense and on its basis — meaning. But in cybernetics, starting from the experience of the creation of technical systems of communication and control, in the notion of "the information" its abstract (one-sided and poor) are fixed — the variety and the complexity of the bearer and the state of the process of the communication of signals, that is syntactic aspects of the information, while the pragmatic and the semantic are ignored, as well as the structural characteristic, which is also not taken as significant by the measure of the quantity of the information. The absolutisation of the syntactic aspect of the information and underestimation of its structural, pragmatic and semantic sides, leads to aparent errors in the process of building models of the animate systems, having psychics.

In the notion of control formal moments are also emphasized, common to the animate and technical systems (for example, any functional dependencies are considered as relations of control).

From all that we can also draw the conclusion about the character and boundaries of the formalisation, mathematisation in cybernetics. For the purpose of rising effectivity and informational value of the notions and methods of cybernetics from the point of view of the mathematical set, we must not only formally describe (in the form of programs) and quantitatively estimate flows, schemes of transmission, reforming the information, embodied in signals. One must also mathematically reflect the organization, structure and functioning of these informational systems. That is why the uncovering, knowing the essence of the selforganizing systems is a key to putting constructively and solving problems of life, psychics, intellect, need oriented activity and emotions of the animate beings. Cybernetic methods and approaches have theoretical and special methodological value for biology, physiology, psychology and complex of sciences on man.

For a more successful advance in the understanding of animate and social systems, their optimum control and also the projection of artificial systems as products of the material and spiritual culture. The improvement of the systems and cybernetic methods is needed. In our view, there are theoretical premises and practical needs for the development of the following tendencies of improvement of these methods.

Firstly, the existence of some independent and "working" variants of the general systems theory, reflecting different sides and approaches to knowing and transforming system-objects, testifies to the need synthesis of these variants of GST into the single general (universal) systems theory. The search for ways of the synthesis (what variant of GST may be the basis of the synthesis, what are the "rational" sides of different variants to preserve, how to link them, etc.), the foundation of models of synthesis etc. are to be the object of a special investigation.

Secondly, the question about the summing of the experience of the application of the general systems and cybernetic methods in their unity in the form of the systems cybernetic method needs solvation. Tendencies of such generalisation and synthesis are already evident in the experience of the so-called systems analysis and in the structure of some variants of GST (for example, in (17)). We have witnessed already the attempts to generalise the experience of the systems approach, GST and cybernetics (see 21).

Thirdly, because cybernetics is not still a theory, organically unifying, linking all of its parts, and because its notions and methods has not yet reached the level of the integral-general notions and methods, there appears the following problem. Further development of cybernetics is connected with the analysis and change in its basis (basic principles, abstractions, notions, premises and directions), the search for a way of transition of its basic notions and methods in integral-general and, at the same time, mathematized notions and methods.

Further investigations in these three directions, in our opinion, will help to enforce special methodological and theoretical functions of the systems theory and cybernetics, their interdisciplinary value.

REFERENCES

1. Kuzmin V. P.: *The Systems Principle in the Theory and Methodology of K. Marx* (Second edition), Moscow 1980.
2. *Investigations in the General Systems Theory,* Moscow 1969.
3. Zelkina O. S.: *Systems-Structural Analysis of the Main Dialectical Categories,* Saratov Univ. Press 1970.
4. Tyukhtin V. S.: *Reflection, Systems, Cybernetics,* Moscow 1972, ch. 1.
5. Mesarović M. D.: "Foundations for a General Systems Theory", in: *General Systems Theory,* Moscow 1966.
6. Boulding K.: "General Systems Theory—the Skeleton of Science", in: *Investigations in General Systems Theory,* Moscow 1969.
7. Simon Th. W.: "Towards an Empiricist Systems Theory", *General Systems,* 1975, No 20.
8. Feigenbaum E. A.: "Artificial Intelligence: Themes in the Second Decade", in: *Cybernetics. Collection of Articles* (New Series), 10 ed., Moscow 1978.
9. Tyukhtin V. S.: "On the Approaches towards the Creation of the General Systems Theory", in: *Systems Analysis and Scientific Knowledge,* Moscow 1978.
10. Lange O.: "Whole and the Development in the Light of Cybernetics", in: *Investigations in the General Systems Theory,* Moscow 1969.

11. **Sengupta S. S.—Ackoff R. L.**: "Systems Theory from an Operations Research Point of View", in: *Investigations in the General Systems Theory*, Moscow 1967.
12. **Sadovsky V. N.**: "General Systems Theory as the Metatheory", *Voprosi filosofii*, 1972, No 4.
13. **Bartalanffy L.**: "General Systems Theory. A Critical Review", in: *Investigations in the General Systems Theory*, Moscow 1969.
14. *Logic and Methodology of Systems Analysis*, Kiev, Odessa 1977, ch. 3—4.
15. **Urmantzev U. A.**: *Symmetry of Nature and the Nature of Symmetry*, Moscow 1974.
16. **Urmantzev U. A.**: "What can give the Biologist the Image of Objects as Systems in the System of Objects of the Same Kind?", *Journal of General Biology*, 1978, vol. 39, No 5.
17. **Bogdanov A. A.**: *General Science of Organization* (*Tectology*), Parts 1—3, Moscow—Leningrad 1925—29.
18. **Mesarović M. D.—Takahara J.**: *General Systems Theory*: *Mathematical Foundations*, N. Y., San Francisco, London 1975.
19. **Mesarović M. D.—Macko D.—Takahara J.**: *Theory of Hierarchical Multilevel Systems*, N. Y., L. 1970.
20. **Zipkin J. Z.**: *Adaptation and Teaching in Automatical Systems*, Moscow 1968.
21. **Van Gigch J. P.**: *Applied General Systems Theory*, N. Y. 1974.

14. The Connection between Natural and Social Sciences in Technological Knowledge

B. S. UKRAINTSEV

The exponential growth of knowledge today is marked by two conflicting trends: the accelerating differentiation of science into individual branches and the tendency to bring together different scientific disciplines through common fundamental ideas. The former reflects the need for the division of scientific labour as the knowledge of the various aspects of development of the inexhaustible objective world is accumulated and deepened. The latter, however, is dictated by the necessities of scientific progress. Science cannot claim this knowledge while turning into a conglomerate of logically ill-connected or disconnected areas, each being terra incognita to the rest.

The problem of the synthesis of knowledge is a historically concrete one and has been solved in different ways in the course of time. Philosophy has been tackling it since ancient times to resolve the question of the relationship between the spirit and the being. In those times the efforts of philosophy were sufficient to somehow bring together the first modest results of scientific studies which, though individual, were for the most part contemplative. Later, however, when individual sciences branched out, the monopoly held by the philosophy of the time on the synthesis of knowledge could not but lead to natural philosophical theories replacing the scientific study of facts by sheer speculation with regard to the structure of the universe.

The philosophy of dialectical materialism completely rejected the claims of natural philosophy that it had created a complete picture of the world. Relying on the results obtained in all sciences, dialectical materialism contributed to the synthesis of knowledge by studying the most general laws governing the development of nature, society and thinking and, thereby, assisted in evolving a general scientific method of studying the objective world. The scientific philosophy of dialectical materialism deliberately limits its participation in synthesizing modern scientific knowledge by providing an ideological and methodological back-up for the solution of this extremely complicated problem.

Although dialectical materialism does envelop extremely wide-ranging philosophical generalizations, regional generalizations are also needed. By this we mean a synthesis of the generalizing ideas permeating many branches of knowledge. They should embrace vast areas of science and logically connect its disciplines by fundamental principles.

The synthesis of knowledge through regional generalizations is a dialectically contradictory process. Synthesis tends to result in the opposite trend, namely the further differentiation of science. For example, some of the generalizing ideas arising from several scientific disciplines, ranging from biology to the humanities, have found their expression in a new scientific trend and new disciplines such as cybernetics engaged in studying the general laws of the functioning of selfgoverning systems, the theory of information, the general theory of systems, and several others.

The technical sciences, which are steadily growing in number, are beginning to play a similar role. They synthesize the results of research not only in the natural but also in the social sciences. This synthesis stems from the nature of the object of study in technical sciences. This object is technology and machinery and the objective laws of its creation, functioning and development.

14.1 THE TECHNOLOGY FACTOR

It is no exaggeration to say that social man and technology have brought about each other's existence and shaped each other as the vital aspects of the material basis of social progress.

K. Marx and F. Engels proved that, in the final analysis, social progress as a whole is determined by the development of the mode of production. This development is in turn stimulated by changes in the productive forces, mainly, improvements in the means of production.

Working implements were the first artificial means created by man to achieve his objectives. The manufacture of such implements marked the beginning of production and the use of technology. The development of working implements and *improvements in the means of production as a whole have been and remain the cornerstone of technological progress.*

Since the nineteenth century the machine designed for the mechanical treatment of the object of labour has been the principal means of labour in large-scale industrial production. It has remained in this capacity for a long time. When studying capitalist production, K. Marx gave his classical definition of this machine as a mechanism, "... which, having received the appropriate motion, performs with its tools the same operations that were earlier performed by the worker with similar tools".[1]

When it was established that the simple working implements of primitive man were the initial phase of technological advancement and that a gradual improvement in the means of production was the nucleus of that advancement, some of the researches studying the technology factor were led to believe that technology in general was identical with working implements and the means of producing material benefits.[2]

Other authors see technology as something more than just a means of production. For example, Yu. S. Meleshchenko believes that "technology is the sum total of things created and used by man through the purposeful utilization of the materials, laws and processes of nature, the things that constitute the material means of his purposeful (labour and, in particular, productive activity".[3]

The author of this definition, underlining the key role of the means of production, sees technology as a material means created by man for his purposeful activity in general. We feel that this view of technology is closer to the actual state of affairs.

Some of the things created by man can play only one part, that is, function as a means of labour. Many other things created by man can easily be employed to achieve not only productive but also non-productive objectives. For example, the telephone, the radio or television set are used as a means of labour in managing a production process. However, the same devices without any change in the nature of their functioning can serve as a means of entertainment in leisure hours.

Evidently, neither the complexity of a device nor the objective for which it has been made (either productive or nonproductive) can serve as a reliable criterion of its falling under the category of technology or of its being excluded from the latter.

These examples certainly give cause for doubting the importance and role of the means of production in technological progress. They have always been and remain the cornerstone of technology, the crucial elements determining its advance. Yet the examples also indicate that technology is not confined to the means of production. It also includes other man-made devices intended to function as the material means of achieving man's non-productive objectives.

It should be pointed out that all artificial devices can possess technological qualities only in interaction with the person using them, who is skilled in employing them as a means of achieving his productive and non-productive objectives. When these devices cease to function and are left on their own, they are no longer part of technology and merely become things of the physical world.

What has been said should not be taken to mean that technology is in need of continuous interaction with man. Present day automatic devices (automatic production lines, shops and even factories, space laboratories) can function independently for a time. However, when they fully lose touch with man, they cease to possess the main feature of technology, their usefulness for achieving the objectives of their maker.

Later, we shall be describing technology as a type of matter and energy organized by man in a special manner (the substance and the field) as well as the methods and skills of using them as an artificial means of achieving the objectives set by society or an individual.

Technology is a social factor, although its material substratum (the substance, the field and energy) has been borrowed from nature. As a social phenomenon, technology has a dialectically contradictory nature. On the one hand, technology is objective in its origin and development in keeping with the objective laws of social progress. It is objective because such are the laws of the motion of matter, the things and energy man has taken from nature and transformed into technical structures.

On the other hand, technology also has a subjective aspect. It is not reduced to natural things in their primordial state. It constitutes substances and forces of nature organized by man's will and programmed to meet his needs. Although the principles and regularities of the organization of natural matter and energy in the technical devices are not arbitrary and, in the final analysis, are determined by the objective laws of the motion of natural and social matter, they do express the relatively independent creative endeavour of man himself. Man selects the necessary combination out of the infinite number of possible actions of the objective laws to use it to achieve his purpose. Man also employs his technical creative effort to transform natural things and to impart a shape and organization to them, which they do not possess in their original state.

By creating technology in the process of labour man directs the motion of natural matter and the currents of energy to achieve the necessary results. In this sense technology is active, yet this is the activity or labour of man.

The interaction of natural and social constituents, the objective and subjective sides of technology determines the specific features of the laws governing the creation, functioning and development of technology and the regularities of technological creative endeavour.

14.2 THE SYSTEMS NATURE OF THE LAWS OF TECHNOLOGY

At all times man has created and used machinery in keeping with the objective laws of nature, society and technology. This should not be taken to mean that the makers of machinery were at all times able to formulate the objective laws of nature as well as those of designing, manufacturing and operating technological devices and structures except for the empirical rules they had discovered which had no theoretical substantiation. The principal laws of technology have not been formulated fully enough even in our age of rapid scientific and technological progress when a large family of technical sciences have appeared.

The laws of technology as well as those of nature, society and thinking were first used in the activities of man before they were given theoretical expression. In other words, they are in the main reflected in the form of empiric knowledge, i.e., the approximate knowledge of the results which might be expected after completing an action in certain conditions, the knowledge of the phenomenon of essence rather than the very essence of things and processes.

Primitive man knew nothing of the theory of the mechanical interaction and the deformation of materials in extremely harsh conditions. However, over the millennia he managed to find such optimal forms of stone tools which amaze specialists in our enlightened age by their perfect geometric shapes that take into account the properties of the material and the objective laws of the mechanical action of these tools. Medieval blacksmiths knew nothing of chemistry or the theory of the hot treatment of metals. Yet the swords of damask steel they made continue to evoke our admiration by their perfection unattainable without empiric knowledge or use of the laws governing the formation of the crystalline structure of steel.

The theoretical analysis and formulation of the objective laws of technology lie within the province of the technical sciences. Our task is that of clarifying the nature of the laws governing technological creative endeavour and the use of technology to the extent to which this problem is linked with the subject of this article.

It is often said in the scientific community that technology is an "applied" natural science while its laws are those of nature used to achieve various objectives. But is this really so?

To answer this question, one must analyse the components forming the objective laws of production, the functioning of the technological means employed to achieve man's objectives.

Let us first turn to the natural substratum of technology. Any technological structure or device is a thing made of substances man takes from nature. All the structures and devices of this type function due to the energy applied to them (coming either from man or derived from other natural sources).

Envidently, therefore, the designing and operation of machinery are based on the objective laws of the motion of matter in general and on those governing the motion of the given type of matter used in the concrete technological device. The natural components of the laws of technology are the objective laws of mechanics (solids, liquids and gases), the laws of thermodynamics, chemical reactions, biological laws, the laws of the formation of the earth's surface and the propagation of rocks, the laws of celestial mechanics and astrophysics, electrodynamics, quantum and nuclear physics, the laws of mental activity, the functioning of self-governing systems, etc.

Every concrete technical device or structure is linked with the action of some of the laws of nature, some of them known and some still unknown. Yet the whole gamut of technical devices, i.e., technology as a whole relies on nearly all of the laws of nature discovered.

As has been said, technology is a social phenomenon and it is not free from the action of the objective laws governing society's development. In a sense, besides the natural matter of technology there is also its "social matter". This can be discovered in the process of its functioning, its interaction with man when it exercises its principal mission of being a means of achieving the objectives of society.

The social components of technology are not so apparent and open to sense perception as the natural ones. If the action of the laws of nature is directly found in the functioning of technology, the action of the social laws is expressed indirectly through relations among people in the course of production.

Not a single technical device or structure can enter the life of society in isolation from its economic laws at the given stage of history. The samples of technology meeting the laws of nature will not become the technology of society if they fail to correspond to its economic conditions. For example, Heron's steam turbine remains a paradox of history because it had no economic foundation in the slave-owning system.

Technology also corresponds to definite legal relations in society. The primitive nature of the crude working implements given to slaves in ancient Rome or any other slave-owning state showed that the direct producers had no rights whatsoever. They had been reduced to the state of draft animals and were not interested in the results of their labour or in keeping those tools in good order. The structure of machines hazardous to the workers exploited during the Industrial Revolution reflected the situation in which capitalist private enterprise was put before their formally declared civil rights. The capitalist regarded the worker as a cheap accessory to the costly machine.

The technology of socialist society is built in keeping with the principles of socialist law. This law regards the working man as the supreme value of society. In accordance with this approach, man is put before technology. He is relieved of hard manual labour and psychological overstrain and every step is taken to ensure his safety, etc.

Technological creative endeavour and the use of technology in a way depend on the ethical principles of society at the given stage in its history. For example, the ethical standards of socialist society prohibit technology from leading to the overexhaustion and humiliation of its users.

People use technology in production, in everyday life, and in their leisure hours. This is why the emotional impact of technology on the human mind, in particular, the aesthetic impact, is of major importance. Technological creative endeavour has been linked with satisfying man's aesthetic needs to varying degrees since ancient times. Much importance was attached to ornamenting weapons and working implements. Technological structures (houses, public buildings, roads, bridges, machines, devices, etc.) have as a rule been made not only with a view to their functional purpose but also to exerting an aesthetic influence on man.

Over the past few decades the first steps have been made to build a technological system to control technology itself. As a result, the laws of natural and artificial languages, semiotics, logic, and psychology are being added to the social components of the technological laws and more actively.

The basic principles of technological creative endeavour, the functioning and development of technology are formed under the influence of the laws of setting and attaining objectives since technology has always been a means of attaining man's aims.

The action of the laws of nature and society acting as components of technological laws is limited to some extent in the course of technological creative endeavour, the functioning and development of technology.

Firstly, the concrete technological structure or device is made and functions in keeping with the many laws governing the motion of matter. These laws limit one another in their action since they lead to opposite results. For example, the thermodynamically rational overheating of steam in the boilers of a power-station is limited by the heat resistance of the boiler steel and the materials of which the turbine rotor is made. The industrial use of thermonuclear energy is limited by the available means of keeping high-temperature plasma within the present volume, etc. For this reason, the laws of technology are the result of a "compromise" between the interacting laws of nature.

Secondly, the laws of nature used in technology limit the action of social laws whereby the means of technology are made and function. For example, the economic efficiency of engines (the cost of fuel per unit of power) is limited by the inevitable heat losses in keeping with the second law of thermodynamics; the profit obtained from transmitting electric power over extremely long distances is limited by the growth of power losses in the transmission lines; the natural contents of a number of technological operations makes it impossible to impart an aesthetic shape to technical structures (blast and open-hearth furnaces, etc.).

Thirdly, social laws, for their part, limit both the use and action of the laws of nature in the technological devices. For example, the use of the most convenient materials as far as their physical and chemical properties are concerned (stainless steel, precious metals, etc.) is limited by economic factors. Many designs of technological devices and structures taking maximum account of the action of the laws of nature are sometimes rejected because they fail to conform to the economic, legal, ethical or aesthetic requirements of society at the given time in history.

In view of their mutual limitation, the natural and social components of the laws of technology never operate independently. Each "component" is in fact an *element* in a system in isolation from which it is no longer a law according to which technology is made and functions.

Unlike the laws of nature and society (reflected by the laws of science in ideal form) each of which can be mentally singled out as allegedly independent and unbreakable into constituents, i.e., an elementary objective law (the law of gravitation, etc.), the laws of technology are a system. Its elements are natural and social components and they lose their qualities as such as soon as we separate them from the system.

The systems nature of the laws of technology is objective in its essence and origin. It is determined by social production and man's inseparable link with nature. This link is expressed in the subordination of natural matter to man as a result of his purposeful activity and the correspondence of the latter to the objective laws of nature and society.

The objective laws of technology have general specific features of their own expressed by the principles of technological creative endeavour and the functioning of technological devices and structures.

As has been pointed out, the principal mission of technology is to act as an artificial means of attaining man's objectives. This circumstance predetermines the general aspect of all the laws of technology expressed by the principle of expediency. Under this principle, all the technological structures or devices, as well as their parts, are made in conformity with the objective, i.e., so that they can serve as a means of achieving man's aims. That is why all the laws of technology are in essence the laws of attaining man's objectives.

Technology exercises its functions in interaction with man who is its user in any conditions. Technology must be subordinated to man. This is why its laws are covered by the general principle of making the possibilities of technology conform to those of the man controlling it or, in other words, the principle of the manageability of technology by man who learns the skills of control during the practical application of technology. A device that man cannot use and is not associated with him as a means of achieving his aims is not part of technology.

Technological creative endeavour is carried out in conformity with the principle of technological applicability. In other words, a new structure must lend itself to manufacturing with the aid of the existing means of production and the available skills as the initial phases of further technological progress.

An important principle of technological creative endeavour and the functioning of technology is reliability. Concrete criteria of reliability were advanced at each historical stage in the technological progress. Until recently this criterion has coincided with that of strength. The sturdier were the buildings, citadels or machine parts, the more reliable they were believed to be. This criterion continues to be valid. Yet it is not applicable to the latest in technology based on electronics, bionic principles, etc. Electronic equipment is not comparable to the body of a rolling mill in strenght. Yet it is expected to function reliably within reasonable limits. What is needed in this case is the preservation of the ability of very sophisticated systems to function when some of their elements break down, and not simply strength or mere resistance.

The laws of technology are also the laws of the efficient functioning of the technological means of achieving man's social and personal objectives. The notion of technological efficiency expresses a relation between the social importance or, in other words, the value of the objective and the social importance of labour and material expenditure in creating the concrete means of technology and their functioning. If the social value of the labour, material and energy expenditure of these means exceeds the social importance of the results of their application as an artificial vehicle of attaining an objective, this technology is considered to be of low efficiency, and society will need a different type of technology in conformity with the efficiency principle.

It must be pointed out that the notion of technological efficiency is broader than the concept of the economic efficiency of technology since the means of technology are often used as a material vehicle in attaining non-productive objectives, for example, ideological ones. In socialist society, for one, the technology of radio broadcasting and television, requiring a vast amount of social and material expenditure, is used for political and cultural enlightenment and education. The efficiency of this technology cannot be expressed in direct economic results.

Large material and labour expenditure on space technology is efficient since our knowledge of the laws of nature and the possibilities of man has improved, although the economic significance of the higher potential of knowledge does not always make itself felt at once or directly.

Last but not least, the laws of technology have another common denominator expressed in the principles of correspondence between technology and society's economic potentialities at the given stage in its history.

Just as one swallow does not make a summer, the individual samples of technological means do not characterize the technical equipment of society. An artificial structure or device becomes part of technology in society when and only when this structure or device turns into a practical means of achieving society's aims.

Each of the principles listed expresses the links existing between the natural and social types of matter in the technology factor, namely, the systems nature of the objective laws of technology.

14.3 TECHNOLOGICAL KNOWLEDGE

The division of knowledge into branches or sciences is as relative as the mental division of the world according to the cardinal points. Yet this division is justified in view of the need to expand and set right the production of knowledge, which is becoming ever more fundamental.

Each branch of knowledge, having turned into an independent science, contains both empiric and theoretical cognizance. Empiric knowledge still prevails in some of the sciences. Others have made big headway in the elaboration of theory. Yet even such a deductive science as mathematics is not completely free of empiric knowledge because its axioms are formulated through the observation of the properties of the objective world and not simply construed.

If we take scientific knowledge as a whole, we can conditionally divide it into two spheres: the field of the scientific knowledge of the laws of nature and society, and that of the laws governing the purposeful transformation of both. The latter includes the knowledge of the laws governing the creation, functioning and development of technology or, in other words, *technological knowledge*. This division is based on the specific features of the science of nature and society, on the one hand, and the peculiarities of the technical sciences, on the other. Let us now consider these features.

The natural and social sciences are engaged in accumulating empiric knowledge and producing theoretical knowledge of the objective laws governing the development of nature, society and thinking. Both set themselves the following principal task: adequately to reflect the objective world, to recognize the way it is in reality, without introducing a subjective element into it. As F. Engels once observed, the problem could only be solved by an infinite series of generations but its fundamental importance for each generation was not diminished because of that.

The production of theoretical knowledge in the natural and social sciences is effected through the discovery and formulation of the objective laws of nature and society, the discovery of trends in the development of things and phenomena as well as predicting the consequences of these trends irrespective of the consciousness and will of man. To transform the world consciously by his activity, man should know its structure and the laws governing its development. It is the natural and social sciences that give him this knowledge.

As a unity of empiric and theoretical knowledge, the technical sciences emerged not so long ago in response to the social requirements of growing industrial production. At first, theoretical technological knowledge was identical to that of the natural sciences, and it was also production that gave impetus to the development of the latter. However, differences later appeared between theoretical and technological mechanics (the theory of mechanisms, the strength of materials, etc.), between theoretical thermodynamics and the theory of the calculation of boilers, heat engines, etc., between physics as a whole and such technical disciplines as electrical and radio engineering, etc.

Since theoretical knowledge in the sphere of the social sciences was insufficiently developed and narrowed by the class framework before Marxism came on the scene, this made it impossible to interpret theoretically the social components of the laws of technology. As a result, the view that technological knowledge was the "applied" knowledge of the laws of nature became even more entrenched.

Technical sciences rely on the empirics and theory of the natural and social sciences. They use the knowledge of their laws. Yet the task of the technical sciences is *that of recognizing the laws of transforming matter in the interests of man.*

As a matter of fact, the technical sciences rely on their own empirics in many ways, first of all, on industrial experience and experimental technological research. Moreover, in view of the advance of the technical sciences their empirical base assumes an importance vital for science as a whole because the practical transformation of nature yields fundamental data on the laws of its development.

The modern technical sciences are coming to rely on a theory of their own more and more. Theoretical technological knowledge substantially differs from that of the natural or social sciences both in character and content.

Firstly, unlike the natural and social sciences, producing the theoretical knowledge of the being, the technical sciences largely produce a theoretical knowledge of the laws and methods of transforming matter (the being) in conformity with the needs of society and the laws of its development.

Secondly, a theoretical knowledge of the technical sciences means possessing a knowledge of the laws governing the purposeful functioning of things and processes, the functioning that does not arise naturally but is necessary to man as an artificial means of attaining his objective.

Thirdly, a theoretical knowledge of the technical sciences is a knowledge of the regularities governing the methods of mutual limitation of the actions of the objective laws of nature and society so that man can define its result.

Fourthly, this knowledge means knowing a number of the regularities expressed in the generalized rules of technological creative endeavour.

Fifthly, theoretical technological knowledge bears the traits of art in the broad sense of the word.

When asked how he made his masterpieces, the famous French sculptor Rodin, is said to have replied "I take a piece of marble and cut off everything superfluous." In a way, technological creative endeavour can be likened to sculpturing. The engineer also takes part of natural matter and "cuts off everything superfluous" to direct its motion into the necessary channel.

Both the natural and social scientists believe that their supreme achievement is the discovery of some regularity in this seemingly chaotic movement of phenomena, the mental clearing of this regularity from any outside influences, the theoretical analysis and formulation of an objective law and the prediction of the results of its action.

While the theoretical knowledge of the natural and social sciences seeks to get rid of the touch of subjectivity, the theoretical knowledge of the technical sciences chooses as its subject the laws governing the interaction between man and nature, the transformation of nature by the will of man and the turning of natural matter and energy into an artificial means of achieving the aims of social man.

When a scientist is unravelling the mystery of nature or society, an objective law of their motion and development, he is performing an immense creative exploit. Yet another scientist is performing a no less magnificent exploit when he discovers the possibilities of putting to practical use the objective laws of nature or society learned. For this utilisation requires its own theory, methods of research, and creative imagination.

As a result, the two types of discovery began to intertwine closely in view of the wide use of experimental research in natural science. A well-arranged experiment is the result of natural scientific theoretical research and technological creative endeavour since every instrument is a technical device acting as an artificial means of cognizing nature.

14.4 THE SYNTHETIC NATURE OF TECHNOLOGICAL KNOWLEDGE

The creation of machinery, its structure and the methods of using it are based on mankind's experience in productive and other activity and rely on a knowledge of society as a whole. Technological knowledge had a complex structure at every stage in its development. It is synthetic by nature.[4] Empiric technological knowledge was already synthetic back in ancient times. The generations of makers of primitive machinery, groping their way through numerous setbacks and losses, accumulated the grains of knowledge of man's actions in the definite natural and social conditions that led to the appearance of acceptable artificial means of producing the material benefits vital for the survival and growth of small communities.

This knowledge included such things as ways of choosing and working natural materials to turn them into tools, a knowing of the probable consequences of various natural phenomena, the nature and sequence of operations in making and using various technical means, the limit of labour expenditure on the production of machinery beyond which the latter does not pay off and, finally, a knowledge of the rituals which evidently served as a rhythm-setting factor in the original technological processes.

It was theoretical research in natural science that provided the impetus to the production of theoretical technological knowledge. First of all, this applies to the study of the laws of mechanical motion.

Theoretical technological knowledge has been synthetic since the time of its appearance because it was impossible to make machinery by directly applying the abstract theory of the mechanical motion of points, etc., without "humanising" it beforehand, without limiting it by those real cases of mechanical motion which lead to results strictly defined in advance in view of their orderly and purposeful nature.

For example, the theory of mechanisms introduces the notion of kinematic pairs as the main elements, tools or parts of artificial mechanisms with a clearly "anthropomorphic", applied character (levers, pulleys, shafts, rack gearings, wheels, etc.). The mechanical theory of the strength of materials of which machinery is made chiefly operates the same standards of strength and signs of destruction as well as geometric forms (beams on various supports, bars, shafts, trusses, etc.) which are imparted by man to the elements of technical structures and devices.

Technological mechanics in all its modifications (the theory of mechanisms, machine parts, construction mechanics, and the "strength of materials") is a theory taking into account not only the least possible expenditure of energy and materials with the given geometric forms of technical structures. It also deals with the economic relations of society at the given stage in its development and the economic effect that could be produced by choosing various shapes and dimensions of technical devices.

It is impossible to discover the direct influences of economic factors on these disciplines since the economic theory is not an organic component of the theory of technological mechanics or the "strength of materials". However, these factors are taken into account indirectly since in each case the theory, considers the optimal versions of the least possible expenditure of materials energy and time in manufacturing machinery and also deals with the historically established mode of operation of the technological means at the given stage.

As the theory and differentiation of the natural sciences keep developing, theoretical technological knowledge is coming to use nearly all the theoretical achievements of natural science as a whole in a transformed shape. This circumstance alone contributes to complicating the structure of technological knowledge and to the appearance of the growing family of technical sciences. At the same time, the penetration of the theory of natural sciences into technological knowledge intensifies the infiltration of social components into the theory of technology.

Theoretical research in the field of thermodynamics and electrodynamics has brought into life such technical sciences as heat and electrical engineering. The new theoretical technological knowledge has served as a basis for a change-over from the mechanical sources of energy to heat and electricity in line with the tendency for production to become increasingly socialized. As a result, it has become possible to introduce new types of energy machinery and to use them on a society-wide scale. Theoretical technological knowledge therefore has had to take into account more and more the action of economic laws and political factors (in setting up international transport, communications, and centralized power supply systems, etc.).

Empiric social components in the theory of technological knowledge are being gradually replaced by the theoretical generalization of the social sciences. The theory of some of the social sciences is merging, as it were, with that of a number of natural sciences in the crucible of theoretical, technological knowledge.

While the first technical sciences were compelled to rely on the theory of the natural sciences alone and outwardly resembled "applied natural science", technical sciences have appeared in recent times in view of the intensification of theoretical research in social science, and are coming to assimilate the results of this research more and more.

The family of sciences constituting a transition from social sciences to technological knowledge and resembling "applied technological social science" is growing rapidly. This applies above all to the intensive penetration of economic ideas and theories into technological knowledge and, on the other hand, the "technicalization" of economic science. The latter development leads to the appearance of such disciplines as the "economics" of various concrete applications of technology (the "economics" of various types of production, transport, etc.).

Such sciences, which are seemingly far apart from each other, like linguistics, the theory of automation, and the theory of computers have begun to come closer together. This trend has produced structural linguistics and the further development of semiotics.

A new technical science, technological aesthetics, has begun to take shape. Its task is to lay a theoretical basis for combining the high functional and aesthetic qualities of new samples of machinery.

A new, independent science, engineering psychology, has come into existence. It lies on the borderline of the technical sciences and those studying the mental peculiarities and possibilities of man.

There is a reason to believe that this is just the beginning of the synthesis of the knowledge of the natural and social sciences in theoretical technological knowledge. This synthesis stems from the systems nature of the objective laws of making, using and developing machinery.

The synthesis of knowledge is causing the opposite tendency to arise, namely, the further differentiation of science. The latter is expressed in the growing number of technical sciences, although the family thereof is already large. This development, in turn, requires regional generalizations, which may well be brought together to form a general theory of technological creative endeavour.

Consequently, we observe two conflicting tendencies here: the differentiation and synthesis of various sciences characterizing the exponential growth of knowledge in our time.

REFERENCES

[1] K. Marx—F. Engels: *Collected Works*, vol. 23, p. 385.

[2] See: A. Yu. Golyan-Nikolsky: *A History of Technology*, part I, Kiev, 1953, p. 4; K. V. Ostrovityanov: "On the Subject of Political Economy", *Kommunist*, 1954, No 4, p. 96; G. V. Osipov: *Technology and Social Progress*, Moscow, 1959, p. 95; G. I. Semenov: *Some of the Methodological Problems of Technical Sciences*, Moscow, 1969, p. 17.

[3] Yu. S. Meleshchenko: *Technological Progress and its Regularities*, Leningrad, 1967, p. 39; Ya. S. Stul: *On the Dialectics of the Notion of Technology*, Chelyabinsk, 1966, p. 21.

[4] See G. I. Semenov: *Some of the Methodological Problems of the Technical Sciences*, p. 33.